JACARANDA

GEOGRAPHY ALIVE 7

AUSTRALIAN CURRICULUM | SECOND EDITION

JACARANDA
GEOGRAPHY ALIVE 7

AUSTRALIAN CURRICULUM | SECOND EDITION

COORDINATING AUTHOR

JUDY MRAZ

CONTRIBUTING AUTHORS

JUDY MRAZ

ANNE DEMPSTER

KATHRYN GIBSON

CATHY BEDSON

TERRY MCMEEKIN

NIRANJAN CASINADER

CLEO WESTHORPE

ALEX ROSSIMEL

JEANA KRIEWALDT

jacaranda
A Wiley Brand

Second edition published 2018 by
John Wiley & Sons Australia, Ltd
42 McDougall Street, Milton, Qld 4064

First edition published 2013

Typeset in 11/14 pt Times LT Std Roman

© John Wiley & Sons Australia, Ltd 2018

The moral rights of the authors have been asserted.

National Library of Australia
Cataloguing-in-publication data

Author:	Mraz, Judy, author.
Title:	Jacaranda Geography Alive 7 for the Australian Curriculum / Judy Mraz, Anne Dempster, Kathryn Gibson, Cathy Bedson, Terry Mcmeekin, Niranjan Casinader, Cleo Westhorpe, Alex Rossimel, Jeana Kriewaldt.
Edition:	Second edition.
ISBN:	9780730346265 (paperback)
Notes:	Includes index.
Target Audience:	For secondary school age.
Subjects:	Geography — Textbooks. Geography — Study and teaching (Secondary) — Australia. Education — Curricula — Australia.
Other Creators/ Contributors:	Dempster, Anne, author. Gibson, Kathryn, author. Bedson, Cathy, author. McMeekin, Terry, author. Casinader, Niranjan, author. Westhorpe, Cleo, author. Rossimel, Alex, author. Kriewaldt, Jeana, author.

Trademarks

Jacaranda, the JacPLUS logo, the learnON, assessON and studyON logos, Wiley and the Wiley logo, and any related trade dress are trademarks or registered trademarks of John Wiley & Sons Inc. and/or its affiliates in the United States, Australia and in other countries, and may not be used without written permission. All other trademarks are the property of their respective owners.

Front cover image: SIBSA Digital Pvt. Ltd. / Alamy Stock Photo

Cartography by Spatial Vision, Melbourne and MAPgraphics Pty Ltd, Brisbane

Illustrated by various artists and the Wiley Art Studio.

Typeset in India by diacriTech

This textbook contains images of Indigenous people who are, or may be, deceased. The publisher appreciates that this inclusion may distress some Indigenous communities. These images have been included so that the young multicultural audience for this book can better appreciate specific aspects of Indigenous history and experience.

It is recommended that teachers should first preview resources on Indigenous topics in relation to their suitability for the class level or situation. It is also suggested that Indigenous parents or community members be invited to help assess the resources to be shown to Indigenous children. At all times the guidelines laid down by the relevant educational authorities should be followed.

Printed in Singapore
M081101R8_210922

CONTENTS

5 Blow wind, blow   101

6 Fieldwork inquiry: What is the water quality of my local catchment?   135

UNIT 2   PLACE AND LIVEABILITY   139

7 Where do Australians live?   141

HOW TO USE

the *Jacaranda Geography Alive* resource suite

For more effective learning, the *Jacaranda Geography Alive* series is now available on the learnON platform. The features described here show how you can use *Jacaranda Geography Alive* to optimise your learning experience.

'Geographical concepts' is a valuable reference section that covers each of the seven concepts.

'Each concept is clearly defined.'

A variety of useful resources support the explanations.

A series of activities to build and develop your understanding of each concept is provided.

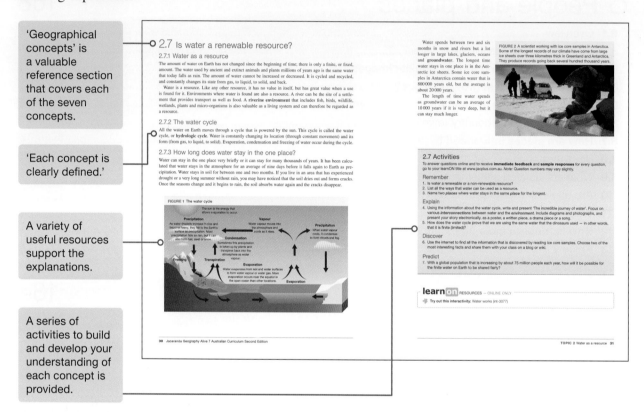

A wide range of engaging and informative visuals are included.

Activities provide you with an opportunity to apply all of the seven concepts.

Linking to *myWorld Atlas* will deepen your understanding.

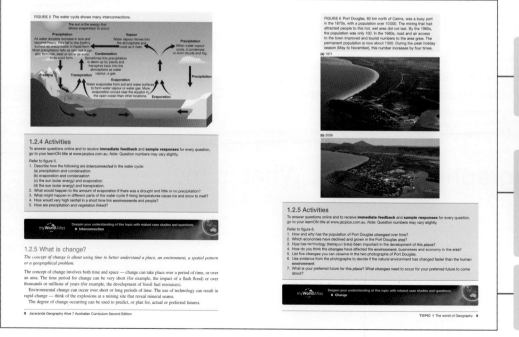

A thought-provoking topic opener sets the scene for your inquiry.

Questions raise issues, link the unit to your life, and prompt you to think about what you already know and feel about the unit.

Evocative and informative images stimulate interest and discussion.

A sequence for your inquiry.

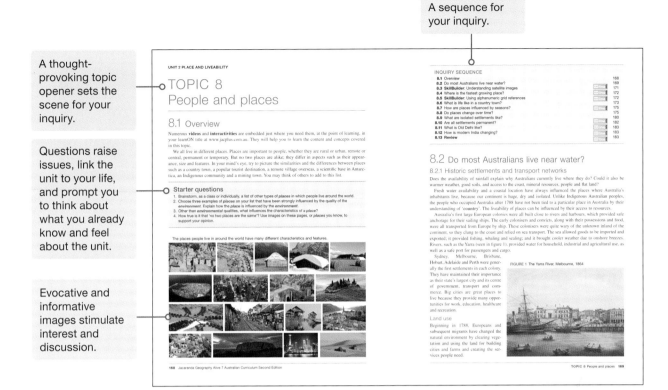

Easily identifiable visual material is referenced in the text and in activities.

Each section begins with a clearly identifiable subtopic number and inquiry question.

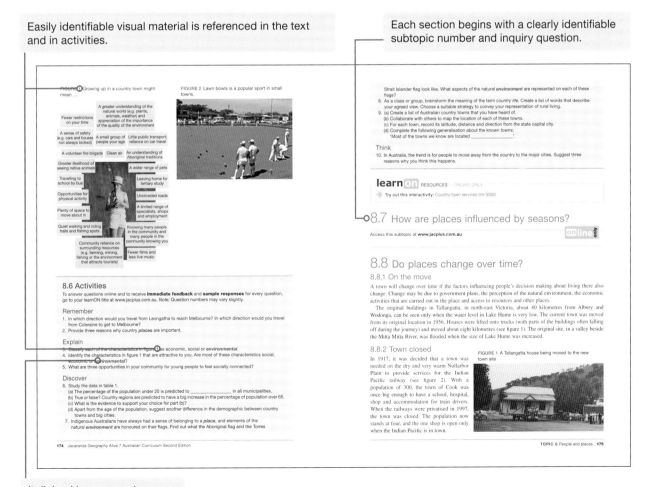

Italicised key concepts are applied to the activities.

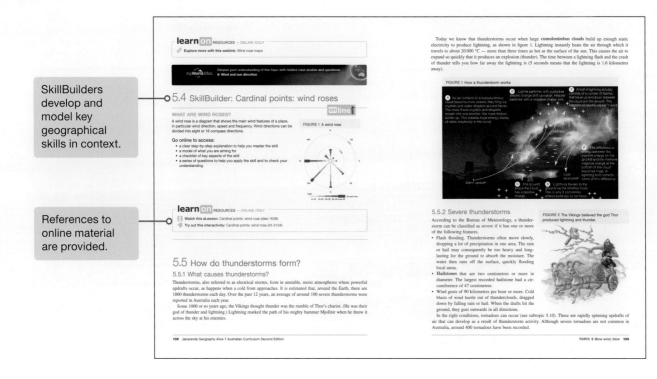

Inside your *Jacaranda Geography Alive learnON*

Jacaranda Geography Alive learnON is an immersive digital learning platform that enables real-time learning through peer-to-peer connections, complete visibility and immediate feedback. It includes:
- a wide variety of embedded videos and interactivities to engage the learner and bring ideas to life
- the **Capabilities** of the Australian Curriculum, available in and throughout the course in activities and **Discussion** widgets
- links to the *myWorld Atlas* for media-rich case studies
- sample responses and immediate feedback for every question
- **SkillBuilders** that present a step-by-step approach to each skill, where each skill is defined and its importance clearly explained
- collaborative activities
- and much more.

ACKNOWLEDGEMENTS

The authors and publisher would like to thank the following copyright holders, organisations and individuals for their assistance and for permission to reproduce copyright material in this book.

Images

• AAP Newswire: **86** (top)/Jenny Evans; **112** (bottom)/Dave Hunt; **117** (middle)/New Zealand Defense Force via AP; **117** (bottom)/New Zealand Defense Force; **194** (middle) • AIATSIS: **161**/This map attempts to represent the language, social or nation groups of Aboriginal Australia. It shows only the general locations of larger groupings of people which may include clans, dialects or individual languages in a group. It used published resources from 1988–1994 and is not intended to be exact, nor the boundaries fixed. It is not suitable for native title or other land claims. David R Horton (creator), © AIATSIS, 1996. No reproduction without permission. To purchase a print version visit: www.aiatsis.ashop.com.au/ • Alamy Australia Pty Ltd: **2** (middle b)/© geogphotos; **10**/ton koene; **13**/Marc Anderson; **22** (top)/robertharding; **28** (e)/Aurora Photos; **31**/ARCTIC IMAGES; **33** (bottom)/Outback Australia; **38** (top)/David Wall; **57** (middle c)/Peter Abell; **112** (top)/model10; **122** (top)/epa european pressphoto agency b.v.; **128** (bottom)/NG Images; **130** (middle)/Michael Dwyer; **143** (top b)/Ian Dagnall; **143** (middle c)/Eric Nathan; **143** (bottom e)/Guillem Lopez; **157** (bottom)/Pulsar Images; **168** (a)/Ian Nellist; **174** (top right)/AA World Travel Library; **181**/© imagebroker / Alamy; **185** (right); **186** (a)/Blend Images; **186** (b)/© Folio Images; **186** (c)/© Bill Bachmann; **196** (b)/© Irene Abdou; **196** (C)/guatebrian; **196** (d)/Friedrich Stark; **196** (e)/Johnny Greig Int; **196** (f)/Piti Anchalee; **200** (c)/ZUMA Press, Inc.; **202** (middle left)/© Geoff Smith; **202** (middle right)/© Thomas Cockrem • Alamy Stock Photo: **17**/Hemis • Andrew Treloar: **47** (middle)/Bureau of Meteorology • Angela Edmonds: **2** (top) • Bureau of Meteorology: **110** (top), **110** (middle), **110** (bottom)/Redrawn from material published by the Bureau of Meteorology; **190** • Creative Commons: **9** (top)/© National Archives of Australia Aerial view of Port Douglas, Queensland | publication-date=**1971** | url=http://trove.nla.gov.au/work/161474801 | accessdate=**9** June 2016; **24** (middle)/Data derived from Department of Industry publication; **25** (middle)/Licensed from the Commonwealth of Australia under a Creative Commons Attribution **3.0** Australia Licence. The Commonwealth of Australia does not necessarily endorse the content of this publication.; **27**/Licensed from the Commonwealth of Australia under a Creative Commons Attribution**3.0** Australia Licence. The Commonwealth of Australia does not necessarily endorse the content of this publication.; **29**/© Crown Copyright 2015 Ministry of Economic Development 2012; **32**/Mattinbgn; **82**/© Australian Bureau of Meteorology; **127** (bottom)/http://en.wikipedia.org/wiki/Avalanche; **132** (top), **132** (bottom), **133** (top)/© State of Queensland Queensland Fire and Emergency Services 2015. Licence at https://creativecommons.org/licenses/by/**4.0**/deed.en.; **146**; **149** (bottom)/© Commonwealth of Australia National Archives of Australia 2015.; **171**/Based on 2015 ABS Statistics; **178**/© The State of Victoria Department of Environment, Land, Water and Planning 2015; **199**/Department of Foreign Affairs and Trade website – www.dfat.gov.au; **199** (a)/© AusAID Department of Foreign Affairs and Trade • ECCA Nepal: **200** (middle a) • European Commission: **116**/c European Union, 2016. Map created by EC-JRC. • Getty Images: **142** (a)/adamkaz; **143** (bottom d)/vicm • Getty Images Australia: **9** (middle), **178** (top left)/Peter Harrison; **38** (middle)/TED MEAD; **45** (bottom)/DEA / PUBBLI AER FOTO; **46** (bottom)/M Swiet Production; **47** (bottom)/Dan Kamminga; **54**/Peter Walton Photography; **57** (bottom)/Paul Chesley; **58**/Karen Kasmauski; **67**/TIMUR MATAHARI; **70**/TORSTEN BLACKWOOD; **75**/Flickr / Julie Fletcher; **92** (middle)/AFP; **96** (bottom)/VANDERLEI ALMEIDA; **99**/MUNIR UZ ZAMAN; **115** (bottom)/Gary Williams / Liaison; **121**/SAM YEH / AFP; **133** (middle right)/dlewis33; **135**/Martin Shields; **141**/btrenkel; **163** (a)/Martin Cohen Wild About Australia; **176** (middle)/Lonely Planet Images / David Wall Photo; **199** (b) • Global Footprint Network: **192**/Global Footprint Network National Footprint Accounts, 2017 Edition Downloaded [12/07/2017] from http://data.footprintnetwork.org. • Grant Gibbs: **200** (middle b)/www.hipporoller.org • International Energy Agency: **25** (bottom)/© OECD/IEA 2015 Key World Energy Statistics, IEA Publishing. Licence: www.iea.org/t&c; as modified by John Wiley & Sons Australia. • Johnny Haglund: **47** (top) • Karen Bowden: **2** (middle a) • MAPgraphics: **22** (middle), **79**; **78, 81**; **125** (top)/US Geological Survey; **150, 170** • Margaret River & Districts: **177**/Margaret River & Districts Historical Society Inc. • NASA: **115** (middle right), **122** (bottom), **129** (middle right) • NASA - JPL: **93** (a bottom), **94** (b)/GSFC/METI/ERSDAC/JAROS, and U.S./Japan ASTER Science Team • NASA Earth Observatory: **128** (middle left) • National Geographic: **184**/Randy Olson • Newspix: **24** (bottom)/Mark Calleja; **28** (f)/Chris Hyde; **92** (bottom)/David Martinelli; **95**/© News Ltd; **97**/David Kapernick; **153** (bottom)/Bob Barker; **186** (f)/News Ltd / Phil Hillyard • Niranjan Casinader: **145** (b), **145** (c), **146**, **157** (b), **158** (a), **158** (a), **158** (b), **159** (b), **159** (c), **159** (d), **159** (e), **159** (f) • NOAA: **129** (bottom)/Climate.gov • Out of Copyright: **169**, **175**/Pictures Collection, State Library of Victoria • Oz Aerial Photography: **176** (bottom) • Photodisc: **28** (a), **28** (c) • Queensland Bulk Water Supply Authority Trading as Seqwater: **88** • Sash Whitehead: **109** (middle) • Shutterstock: **1**/Scanrail1; **1** (a)/Honza Krej; **1** (b)/john austin; **1** (c)/Janelle Lugge; **1** (d)/Willyam Bradberry; **1** (e)/Lucian Coman; **1** (f)/Adisa; **1** (g)/think-4photop; **1** (h)/Alexander Smulskiy; **5** (middle)/Christian Draghici; **11**/Doidam 10; **18** (a)/Monkey Business Images; **18** (b)/freya-photographer; **18** (c)/WitthayaP; **18** (d)/Catalin Petolea; **20**/18042011; **25** (top)/© Dmitriy Kuzmichev; **28** (d)/Sky Light Pictures; **36** (right)/Keith Wheatley; **41**/Johan Swanepoel; **57** (middle a)/grafvision; **57** (middle b)/kezza; **68** (bottom), **136**/goodluz; **76** (a)/Nils Versemann; **76** (b)/Martin Valigursky; **76** (c)/Jiratsung; **90**/zstock; **98** (middle), **157** (a)/yampi; **111** (botttom)/Jill Battaglia; **111** (middle)/Ryszard Stelmachowicz; **125** (bottom)/Matt Jeppson; **128** (middle right)/© joyfull; **131** (top)/Toranico; **68** (bottom), **136**/Goodluz; **142** (b)/Anton_Ivanov; **143** (a)/T photography; **148** (a)/qingqing; **148** (b)/ Ivonne Wierink; **148** (c)/NigelSpiers; **148** (d)/AJP; **164** (b)/sigurcamp; **164** (c)/Production Perig; **165** (d)/Dan Breckwoldt; **165** (e)/LingHK; **168** (b)/haveseen; **168** (c)/Nickolay Vinokurov; **168** (d)/Alexander Chaikin; **168** (e)/Anki Hoglund; **168** (f)/ Mohamed Shareef; **168** (g)/Andrey Bayda; **168** (h)/stocker**1970**; **168** (i)/Francesco R. Iacomino; **168** (j)/Jane Rix; **168** (k)/ Aleksandar Todorovic; **168** (l)/Kenneth Dedeu; **168** (m)/JeniFoto; **168** (n)/leoks; **168** (o)/StanislavBeloglazov; **168** (p)/ chuyu; **168** (q)/Vladimir Melnik; **185** (left)/Sveta Yaroshuk; **186** (d)/Blend Images; **186** (e)/© studioflara; **186** (g)/© Martin Allinger; **202** (bottom)/pbk-pg; **204** (bottom)/Monkey Business Images; **205**/monticello; **208**/Joao Virissimo; **210**/Mark_ and_Anna_Wilson; **210** (top)/kurhan• Spatial Vision: **3** (left)/NASA Earth Observatory; **3** (right)/USAID, FEWS NET 2011; **3** (bottom), **44**/Geophysical Fluid Dynamics Labartory, National Oceanic and Atmospheric Administration; **4**, **37**, **71**, **84**, **105** (a), **114** (bottom), **151** (top), **156**, **163**; **5** (bottom), **6** (top), **12** (bottom), **49**, **56**, **70**, **85**, **86** (bottom), **145**, **188**; **23**/ Department of Environment and Water Resources; **26**/© BP Statistical Review of World Energy 2011 http://www.bp.com/ assets/bp_internet/globalbp/globalbp_uk_english/reports_and_publications/statistical_energy_review_2011/STAGING/ local_assets/pdf/statistical_review_of_world_energy_full_report_**201**; **34**/BGR & UNESCO 2008: Groundwater Resources of the World **1** : 25 000 000. Hannover, Paris. was: data copyright 2012 WHYMAP / Map redrawn by Spatial Vision; **36** (left), **151** (bottom)/Geoscience Australia; **42**/WorldClim; **50** (top), **50** (bottom a), **50** (bottom b); **53**/Mekonnen, M.M. and Hoekstra, A.Y. 2011 National water footprint accounts: the green, blue and grey water footprint of production and consumption, Value of Water Research Report Series No.**50**, UNESCO-IHE, Delft, the Netherlands; **60** (top)/© Copyright World Health Organisation WHO, 2012; **61**/© Data sourced from World Health Organisation 2012 / Redrawn by Spatial Vision; **62**/World Health Organisation / UNICEF Joint Monitoring Program JMP for Water Supply and Sanitation; **89**/Base Data © Copyright Commonweath of Australia Geosicence Australia 2006; **96** (top)/Natural Earth Data; **133** (middle left)/ North Carolina Floodplain Mapping Program; **153** (top)/Mapland - Resource Information A division of: Department for Environment Heritage and Aboriginal Affairs; **176**/Geoscience Australia; **180**, **193**/Natural Earth Data; **196**/Food and Agriculture Organisation • State Library of Victoria: **145** (a)/John T Collins / Walkerville Lime Kilns / Pictures Collection • Stephen Locke: **101** • Water Footprint Network: **65**/WaterStat, The Hague, the Netherlands • Wiley Media Manager: **60** (bottom)/Jake Lyell / Alamy Stock Photo

Text

• © Australian Curriculum, Assessment and Reporting Authority (ACARA) 2010 to present, unless otherwise indicated. This material was downloaded from the Australian Curriculum website (www.australiancurriculum.edu.au) (Website) (accessed September 5, 2017) and was not modified. The material is licensed under CC BY 4.0 (https://creativecommons.org/licenses/ by/4.0). Version updates are tracked on the 'Curriculum version history' page (www.australiancurriculum.edu.au/Home/ CurriculumHistory) of the Australian Curriculum website. • ARCADIS: **191**/Table created based on data from Arcadis publication Sustainable Cities Index 2015 • Bureau of Meteorology: **120** (top) • Creative Commons: **114**/ClearlyExplained. Com; **156** (bottom); **173**/© The State of Victoria Department of Environment, Land, Water and Planning 2015; **177** (middle right)/Data based on ABS statistics • United Nations: **61**

Every effort has been made to trace the ownership of copyright material. Information that will enable the publisher to rectify any error or omission in subsequent reprints will be welcome. In such cases, please contact the Permissions Section of John Wiley & Sons Australia, Ltd.

TOPIC 1
The world of Geography

1.1 Overview

1.1.1 What is Geography?

Numerous **videos** and **interactivities** are embedded just where you need them, at the point of learning, in your learnON title at www.jacplus.com.au. They will help you to learn the content and concepts covered in this topic.

The world around us is made up of a large range of interesting places, people, cultures and environments. Geography is a way of exploring, analysing and understanding this world of ours: especially its people and places. Studying Geography at school allows you to build up your knowledge and understanding of our planet, at different scales: the local area, our nation, our region and our world. In essence, geographers investigate the characteristics of places and the relationships between people and places.

1.1.2 Geography is … about our wonderful world

Have you ever visited a place other than the one you live in? If so, you probably would have noticed some of the features and characteristics are similar, and some are different. Geographers aim to understand these characteristics as well as the relationship between people and the different environments around us.

As a geographer, you answer questions ranging from the local to the global, in the past, present and future. Along the way you will develop skills and inquiry methods to answer these questions for yourself.

FIGURE 1 Our planet is made up of a large variety of fascinating places, peoples, cultures and environments.

1.1.3 Geography is … something you do

One of the best parts of studying Geography is the opportunity to visit places outside the classroom. Going on a field trip allows you to collect data and information for yourself and to work collaboratively with other members of your class.

Geographers use what is called an 'inquiry' approach. This means that you will investigate geographical questions by collecting, analysing, and interpreting information and data in order to develop your own understanding and draw your own conclusions. This helps you develop proposals for what should happen and what action should be taken in the future.

Studying Geography develops a wide range of skills that you can apply in your everyday life, in your future life and possibly in your career.

FIGURE 2 Using maps to work out locations and to plot data

FIGURE 3 Conducting a survey in the field

FIGURE 4 Collecting your own data and information

1.1.4 Geography is … a way of thinking

Geography is a way of thinking and a way of looking at the world. One of the key tools geographers use is a map. If you look really carefully at them, maps (such as the ones on this page and the next) contain a lot of information. As a student you will often use a variety of different types of maps produced by someone else (e.g. from this textbook, atlases and online). However, as a geographer you will produce your own maps and spatial information, by hand or digitally. Using and interpreting maps are important skills you will develop. It is also important to identify major patterns and trends in maps in order to unlock information they contain.

As a geographer you will use a set of geographical concepts to not only help you think geographically but also to investigate and understand the world.

These concepts are space, place, interconnection, change, environment, sustainability and scale (see pages 6–12).

As a geographer you should also ask yourself: 'What can I do and contribute as an informed and responsible citizen to make this world a better place?'.

FIGURE 5 Maps: a key tool for the geographer

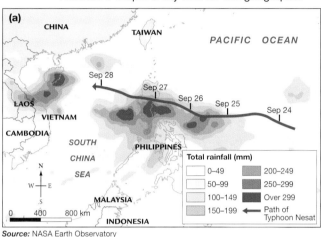

Source: NASA Earth Observatory

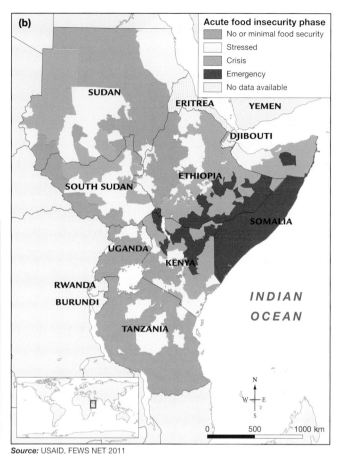

Source: USAID, FEWS NET 2011

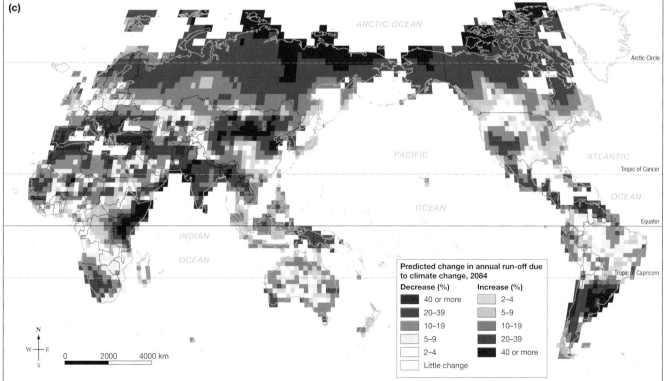

Source: Geophysical Fluid Dynamics Labratory, National Oceanic and Atmospheric Administration

FIGURE 6 Topographic maps are very useful for geographers as they provide a large amount of detail about places and environments.

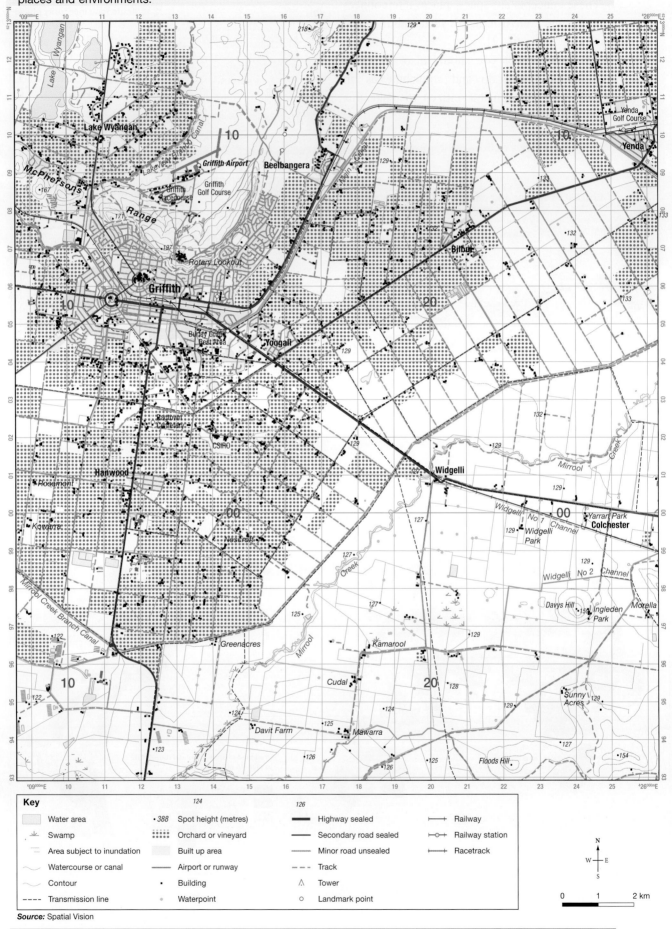

Key

Water area	•388 Spot height (metres)	Highway sealed	Railway
Swamp	Orchard or vineyard	Secondary road sealed	Railway station
Area subject to inundation	Built up area	Minor road unsealed	Racetrack
Watercourse or canal	Airport or runway	Track	
Contour	• Building	∧ Tower	
Transmission line	• Waterpoint	○ Landmark point	

N
W E
S

0   1   2 km

Source: Spatial Vision

1.2 Geographical concepts

1.2.1 Overview

Geographical concepts help you to make sense of your world. By using these concepts you can both investigate and understand the world you live in, and you can use them to try to imagine a different world. The concepts help you to think geographically. There are seven major concepts: *space*, *place*, **interconnection**, *change*, *environment*, *sustainability* and *scale*.

In this book, you will use the seven concepts to investigate two units: *Water in the world* and *Place and liveability*.

1.2.2 What is space?

Everything has a location on the space that is the surface of the Earth, and studying the effects of location, the distribution of things across this space, and how the space is organised and managed by people, helps us to understand why the world is like it is.

A place can be described by its absolute location (latitude and longitude) or its relative location (in what direction and how far it is from another place).

FIGURE 1 A way to remember these seven concepts is to think of the term SPICESS.

FIGURE 2 Australian annual rainfall variability, 1900–2003

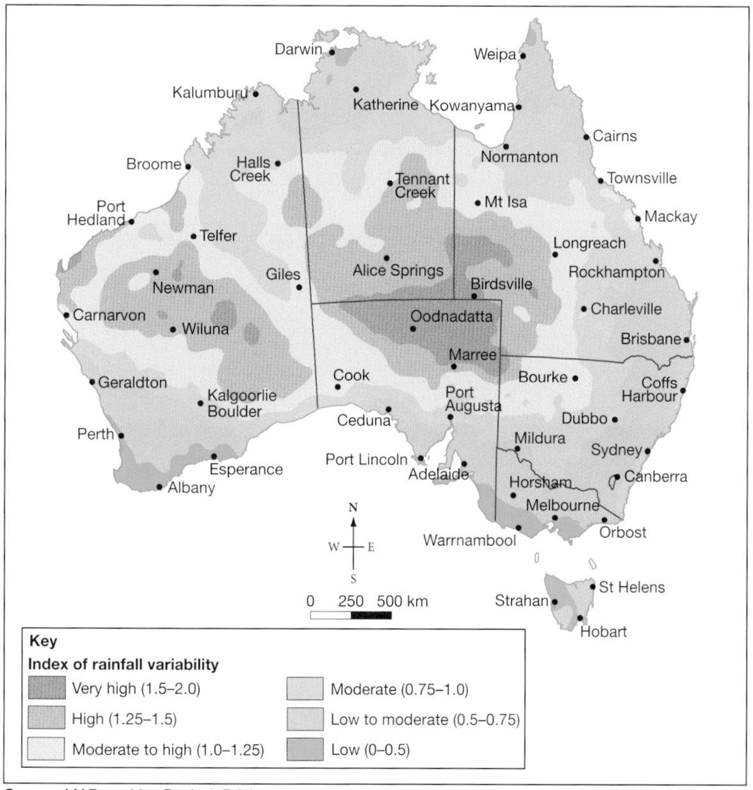

Key
Index of rainfall variability
Very high (1.5–2.0)
High (1.25–1.5)
Moderate to high (1.0–1.25)
Moderate (0.75–1.0)
Low to moderate (0.5–0.75)
Low (0–0.5)

Source: MAPgraphics Pty Ltd, Brisbane

FIGURE 3 The amount of rain that falls in Australia varies from place to place, as this rainfall map shows.

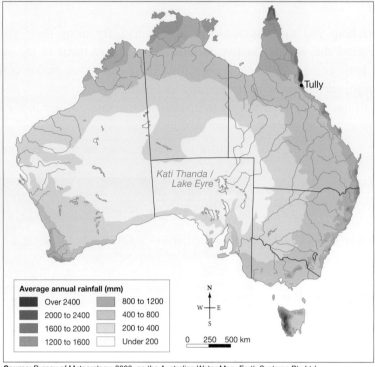

Source: Bureau of Meteorology, 2003, on the Australian Water Map, Earth Systems Pty Ltd

1.2.2 Activities

To answer questions online and to receive **immediate feedback** and **sample responses** for every question, go to your learnON title at www.jacplus.com.au. *Note:* Question numbers may vary slightly.

Refer to figures 2 and 3.
1. Use an atlas to give the absolute location (latitude and longitude) of the capital city of the state/territory in which you live.
2. In which direction and how far is your capital city from Alice Springs (relative location)?
3. Describe the *spatial* distribution of capital cities in Australia.
4. Describe the distribution of rainfall across Australia. Why might one place have more or less rainfall than another?
5. How does rainfall (or lack of rainfall) help explain the distribution of Australia's major cities? What is the relationship between rainfall and population location?
6. Find where you live on the maps. How is the location of your *place* influenced by rainfall and rainfall variability?

 my**World**Atlas   Deepen your understanding of this topic with related case studies and questions.
❯ **Space**

1.2.3 What is place?

The world is made up of places, so to understand our world we need to understand its places by studying their variety, how they influence our lives and how we create and change them.

You often have mental images and perceptions of places — your city, suburb, town or neighbourhood — and these may be very different from someone else's perceptions of the same places.

FIGURE 4 Mount Tom Price township and mine in Western Australia, with fly in, fly out (FIFO) workers huts in the left foreground

1.2.3 Activities

To answer questions online and to receive **immediate feedback** and **sample responses** for every question, go to your learnON title at www.jacplus.com.au. *Note:* Question numbers may vary slightly.

Refer to figure 4.
1. Where is this *place* located? (Refer to an atlas.)
2. What is this *place* like? (What are its natural characteristics? What are its human characteristics?)
3. How have people changed this *place*?
4. Why do you think that Mount Tom Price township was settled in this location?
5. What services and facilities are provided by this *place*? How is this different to where you live?
6. How do you think the *environment* of Mount Tom Price affects the people who live there?
7. How might this *place* change in the future?
8. How do you think this *place* affects the lives of the people who live there?

myWorldAtlas — Deepen your understanding of this topic with related case studies and questions.
❯ **Place**

1.2.4 What is interconnection?

People and things are connected to other people and things in their own and other places, and understanding these connections helps us to understand how and why places are changing.

An event in one location can lead to change in a place some distance away.

FIGURE 5 The water cycle shows many interconnections.

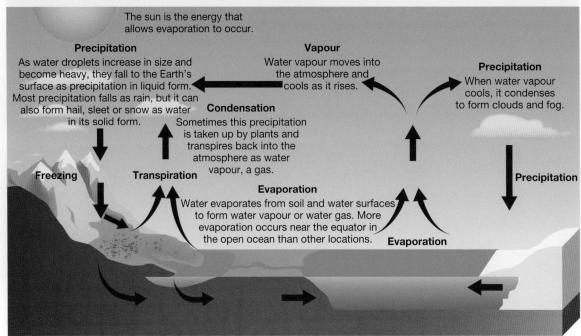

The sun is the energy that allows evaporation to occur.

Precipitation
As water droplets increase in size and become heavy, they fall to the Earth's surface as precipitation in liquid form. Most precipitation falls as rain, but it can also form hail, sleet or snow as water in its solid form.

Vapour
Water vapour moves into the atmosphere and cools as it rises.

Precipitation
When water vapour cools, it condenses to form clouds and fog.

Condensation
Sometimes this precipitation is taken up by plants and transpires back into the atmosphere as water vapour, a gas.

Freezing

Transpiration

Evaporation
Water evaporates from soil and water surfaces to form water vapour or water gas. More evaporation occurs near the equator in the open ocean than other locations.

Evaporation

Precipitation

1.2.4 Activities

To answer questions online and to receive **immediate feedback** and **sample responses** for every question, go to your learnON title at www.jacplus.com.au. *Note:* Question numbers may vary slightly.

Refer to figure 5.
1. Describe how the following are *interconnected* in the water cycle:
   (a) precipitation and condensation
   (b) evaporation and condensation
   (c) the sun (solar energy) and evaporation
   (d) the sun (solar energy) and transpiration.
2. What would happen to the amount of evaporation if there was a drought and little or no precipitation?
3. What might happen in different parts of the water cycle if rising temperatures cause ice and snow to melt?
4. How would very high rainfall in a short time link *environments* and people?
5. How are precipitation and vegetation linked?

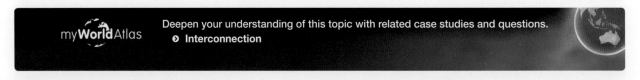

myWorldAtlas
Deepen your understanding of this topic with related case studies and questions.
❯ Interconnection

1.2.5 What is change?

The concept of change is about using time to better understand a place, an environment, a spatial pattern or a geographical problem.

The concept of change involves both time and space — change can take place over a period of time, or over an area. The time period for change can be very short (for example, the impact of a flash flood) or over thousands or millions of years (for example, the development of fossil fuel resources).

Environmental change can occur over short or long periods of time. The use of technology can result in rapid change — think of the explosions at a mining site that reveal mineral seams.

The degree of change occurring can be used to predict, or plan for, actual or preferred futures.

FIGURE 6 Port Douglas, 60 km north of Cairns, was a busy port in the 1870s, with a population over 10 000. The mining that had attracted people to this hot, wet area did not last. By the 1960s, the population was only 100. In the 1980s, road and air access to the town improved and tourist numbers to the area grew. The permanent population is now about 1300. During the peak holiday season (May to November), this number increases by four times.

(a) 1971

(b) 2009

1.2.5 Activities

To answer questions online and to receive **immediate feedback** and **sample responses** for every question, go to your learnON title at www.jacplus.com.au. *Note:* Question numbers may vary slightly.

Refer to figure 6.
1. How and why has the population of Port Douglas *changed* over time?
2. Which economies have declined and grown in the Port Douglas area?
3. How has technology (transport links) been important in the development of this *place*?
4. How do you think the *changes* have affected the *environment*, businesses and economy in the area?
5. List five *changes* you can observe in the two photographs of Port Douglas.
6. Use evidence from the photographs to decide if the natural environment has changed faster than the human *environment*.
7. What is your preferred future for this *place*? What *changes* need to occur for your preferred future to come about?

 Deepen your understanding of this topic with related case studies and questions.
  ❯ Change

1.2.6 What is environment?

People live in and depend on the environment, so it has an important influence on our lives.

The environment, defined as the physical and biological world around us, supports and enriches human and other life by providing raw materials and food, absorbing and recycling wastes, and being a source of enjoyment and inspiration to people.

FIGURE 7 Pacific Islanders use traditional methods to fish sustainably.

1.2.6 Activities

To answer questions online and to receive **immediate feedback** and **sample responses** for every question, go to your learnON title at www.jacplus.com.au. *Note:* Question numbers may vary slightly.

Refer to figure 7.
1. Do you think the photograph of Pacific Islanders fishing is a natural *environment* or a human *environment*? Explain.
2. Does this *environment* appeal to you? Would you like to visit this *place*? Why? Why not?
3. Which resource/s do you think people would obtain from this *environment*?
4. Describe how these people are fishing. Why might this be *sustainable*?
5. List the impacts on this *environment* if a factory was built on the edge of the water.
6. How have people changed this *environment* (for better or worse)? What are the positive and the negative aspects of this?
7. How might technology *change* this *environment* to make it less *sustainable*?

 my**World**Atlas   Deepen your understanding of this topic with related case studies and questions.
❯ **Environment**

1.2.7 What is sustainability?

Sustainability is about maintaining the capacity of the environment to support our lives and those of other living creatures.

Sustainability is about the interconnection between the human and natural world and who gets which resources and where, in relation to conservation of these resources and prevention of environmental damage.

FIGURE 8 The process of shifting cultivation means that farmers move on when an area becomes unproductive, allowing the land to recover.

A clearing is made by cutting and burning vegetation. This is known as 'slash and burn'.
Crops are planted and grow well.
After three to four years the nutrients in the soil have been used up and the crops don't grow as successfully.
The clearing is abandoned as farmers move to a new area. As a result the clearing gradually returns to its natural state.

1.2.7 Activities

To answer questions online and to receive **immediate feedback** and **sample responses** for every question, go to your learnON title at www.jacplus.com.au. *Note:* Question numbers may vary slightly.

Refer to figure 8.

1. What evidence is there that the ***environment*** shown in the image is being conserved?
2. How is this area being maintained so that its resources can be supplied continuously into the future?
3. Is there evidence that the aesthetic (beauty) elements of this ***environment*** are being protected?
4. How would the ***environment*** be changed if all the area shown was cleared and farmed at the same time? Would this be ***sustainable***?
5. Can you think of any farming methods that are not ***sustainable***? List these.

myWorldAtlas

Deepen your understanding of this topic with related case studies and questions.

- **Sustainability**

1.2.8 What is scale?

When we examine geographical questions at different spatial levels we are using the concept of scale to find more complete answers.

Scale can be applied at personal and local levels to regional, national or global levels. Looking at things at a range of scales allows a deeper understanding of geographical issues.

Different factors can be involved in explaining phenomena at different scales. Local events can have global outcomes; for example, removing areas of forest at a local scale can have an impact on climate at a global scale. A policy at a national scale, such as forest protection, can have an impact at a local scale, such as the protection of an endangered species.

FIGURE 9 Mental map of Jayden's local place (a) by Jayden and (b) by Annette, Jayden's mother

FIGURE 10 Railway route and main settlements between Sydney and Perth

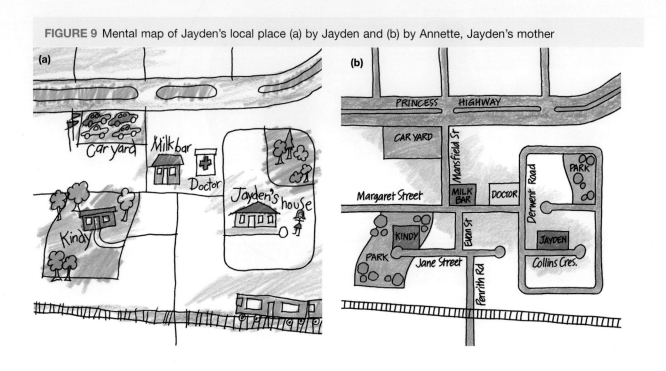

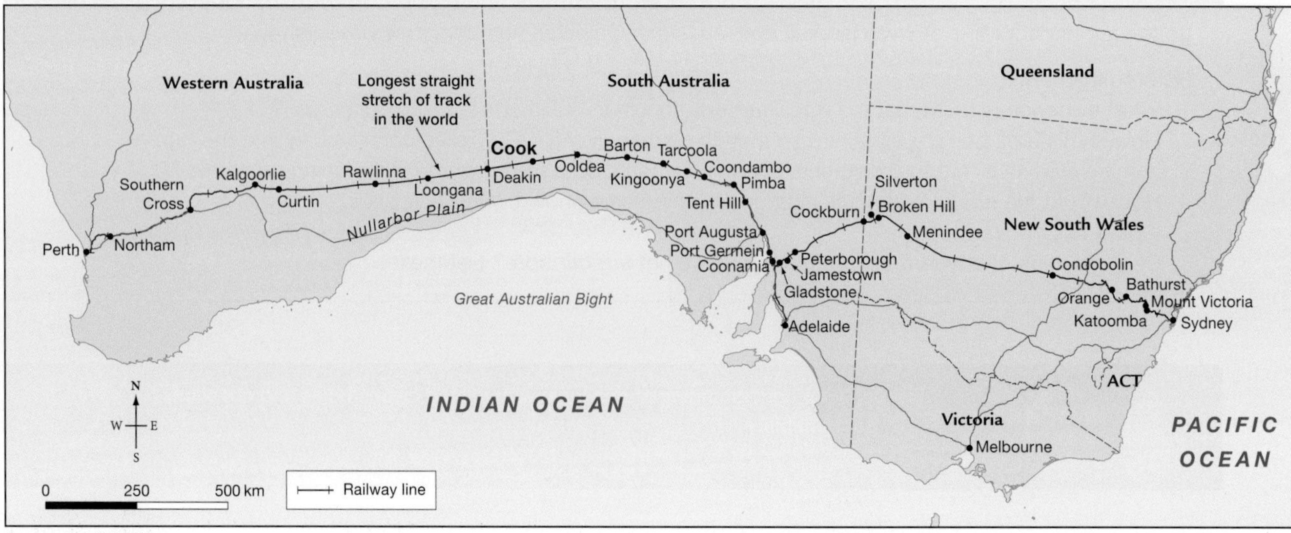

Source: Spatial Vision

1.2.8 Activities

To answer questions online and to receive **immediate feedback** and **sample responses** for every question, go to your learnON title at www.jacplus.com.au. *Note:* Question numbers may vary slightly.

Refer to figures 9 and 10.
1. If you were to zoom in on the areas on these maps, would you see more or less detail?
2. List the detail and information you can see on the railway map. Compare this to the local neighbourhood maps. Which gives you more information?
3. Refer to the railway map. What might be the relationship between the location of settlements and the location of the railway?
4. The railway map is of a regional *scale*. Which region of Australia is it showing?
5. Use the scale to measure the longest straight stretch of railway shown on the map. How long is it? Why is it significant?
6. What is the main information each map is trying to show?

---

 myWorldAtlas   Deepen your understanding of this topic with related case studies and questions.
❯ **Scale**

1.3 Review

1.3.1 Applying the concepts

Port Moresby is the capital of Papua New Guinea and is located on the country's south-eastern coast. With a population of approximately 350 000, the city has a mix of high-rise urbanised landscapes and village landscapes. There are both poor and rich people who live in the city. The formal settlement in this image consists of the buildings and roads in the background, which have been planned. The informal settlement consists of houses on stilts, which have been built over the water, in many cases without permits, on state land. Up to half of Port Moresby's population now lives in these squatter settlements due to a lack of affordable housing.

FIGURE 11 Port Moresby is a mixture of high-rise urbanised landscapes and village landscapes.

Formal settlement
- Street layout planned
- Rubbish collection, power, water and sanitation available
- Cost of housing and services is very high.
- Public transport
- Street lighting
- Public buildings (such as museums) and gardens
- Sealed roads

Informal settlement
- The number of these settlements is growing to meet the needs of increased migration to the city.
- Found materials are sometimes used in housing construction.
- Some houses are built over water to avoid disputes over land.
- Streets unplanned
- Housing does not always withstand heavy rain and wind.
- Poor access to power, water and sanitation
- Many households plant food crops.
- Many dirt roads

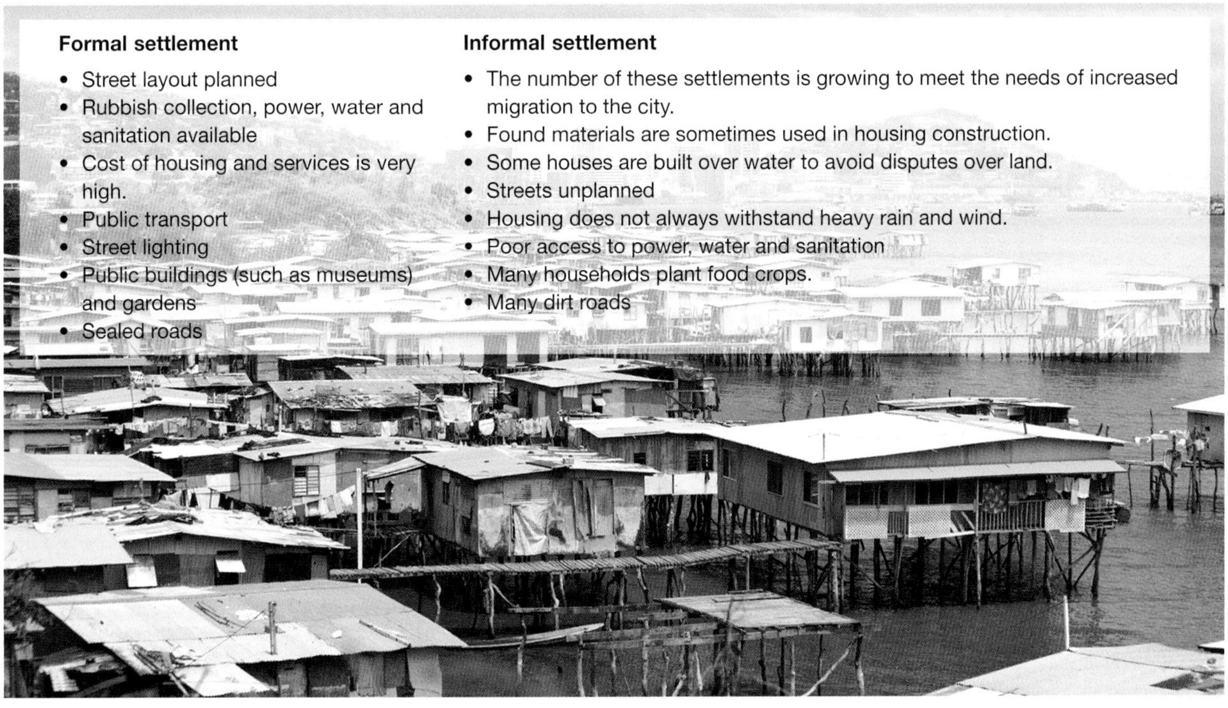

1.3.1 Activities

To answer questions online and to receive **immediate feedback** and **sample responses** for every question, go to your learnON title at www.jacplus.com.au. *Note:* Question numbers may vary slightly.

Refer to figure 11.

1. Where is Port Moresby located? (*space*)
2. What is your perception (feelings) about this *place*?
3. How do you think the people living in the informal settlements might feel about their *place*? How might this compare to those living in the formal settlement?
4. Describe the human and natural characteristics of this environment. (*space*)
5. How has this *environment* been *changed* by people? Is there any evidence of the original natural environment?
6. List the resources that this *environment* provides for people.
7. How has the informal settlement met the needs of the population? (*space*)
8. How does climate affect the informal settlement? (*space*)
9. Describe the *interconnections* between:
   (a) water and buildings in the informal settlement
   (b) the formal and informal settlements.
10. How would people in the informal *environment* obtain their water?
11. Should people be allowed to live in the informal settlement? Is this a *sustainable* use of the local resources?
12. How does the *scale* of the buildings differ in the two settlements? How does this reflect the services each location has access to?
13. Describe five differences between the two settlements. (*scale*)
14. What is your preferred future for this place? What *changes* need to occur for your preferred future to come about?

UNIT 1
WATER IN THE WORLD

Water is all around us — in the atmosphere as clouds; as snow on mountain tops; in rivers, lakes and dams on the Earth's surface; in our swimming pools, on our farms and in our taps. It is one of our most valuable resources, because humans cannot survive without water. Therefore, we need to learn to use and look after this resource very carefully.

Water in its three forms: as a liquid, in the lake; as a solid, in the snow and ice on top of the mountains; and in vapour form, in the clouds

TOPIC 2
Water as a resource

2.1 Overview

Numerous **videos** and **interactivities** are embedded just where you need them, at the point of learning, in your learnON title at www.jacplus.com.au. They will help you to learn the content and concepts covered in this topic.

2.1.1 Introduction

Have you ever stopped to think about the resources you need to survive every day? Fortunately, the Earth supplies us with the natural resources we need for our food, shelter, clothing, and energy for our homes and factories. These resources include water, fossil fuels and mineral deposits. However, access to these supplies is not distributed equally around the planet, and attitudes towards them may differ or change over time. As the global population increases, great damage is being done to the environment as a result of using these resources. Moreover, we need to carefully manage our use of them to ensure that these resources are available for use in the future.

Starter questions

1. List some natural resources that are used by your family on a daily basis.
2. List some of the different ways water is used in our lives.
3. Look around your classroom and list items made from natural resources. In some cases, you may need to explain the links between the objects and the resources used to make them.
4. How can the natural resources that you use harm the *environment*?
5. Discuss how your family tries to reduce waste and protect resources.

Pacific Islanders use traditional methods to fish sustainably.

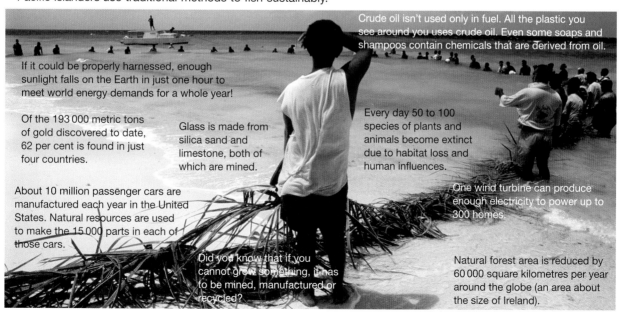

If it could be properly harnessed, enough sunlight falls on the Earth in just one hour to meet world energy demands for a whole year!

Of the 193 000 metric tons of gold discovered to date, 62 per cent is found in just four countries.

About 10 million passenger cars are manufactured each year in the United States. Natural resources are used to make the 15 000 parts in each of those cars.

Glass is made from silica sand and limestone, both of which are mined.

Did you know that if you cannot grow something, it has to be mined, manufactured or recycled?

Crude oil isn't used only in fuel. All the plastic you see around you uses crude oil. Even some soaps and shampoos contain chemicals that are derived from oil.

Every day 50 to 100 species of plants and animals become extinct due to habitat loss and human influences.

One wind turbine can produce enough electricity to power up to 300 homes.

Natural forest area is reduced by 60 000 square kilometres per year around the globe (an area about the size of Ireland).

2.2 What is a resource?

2.2.1 Why do we need resources?

We depend on natural resources to survive. We need water to drink, soil to produce our food, and forests and mines to supply other materials. Natural resources are raw materials that occur in the environment and which are necessary or useful to people. They include soil, water, mineral deposits, **fossil fuels**, plants and animals.

Think about all the resources you have used today from the time you woke up until the time you reached the school gate. Perhaps you used water to shower, brush your teeth, wash the dishes or as a refreshing drink? Consider all the different foods that had to be farmed to provide the ingredients for your breakfast. Finally, how did you get to school? If you used a form of transport, there is a good chance a resource powered it!

There are two types of natural resources: non-renewable and renewable.

Renewable resources are those that can be replaced in a short time. For example, solar energy is a renewable resource that can be used for heating water or generating electricity. It is never used up and is constantly being replaced by the sun.

FIGURE 1 Many resources are required to provide a family with breakfast.

FIGURE 2 Natural resources — renewable and non-renewable

```
                                                    Recyclable
  Consumed by use                                   Aluminium, gold,
  Fossil fuels, coal,                               lead, silver, tin,
  oil, natural gas                                  mercury

                        Non-renewable
                        Replaced over
                        millions of
                        years

                        Natural resources

                        Renewable
                        Replaced over
                        short period
                        of time
                                                    Supply can be
  Constant supply                                   affected by
  Solar energy,                                     human activity
  tides, waves,                                     Fish, forests, soil,
  water, air                                        water, under-
                                                    ground water
```

Non-renewable resources are those that cannot be replaced in a short time. For example, fossil fuels such as oil, coal and natural gas are non-renewable because they take thousands of years to be replaced.

We cannot make more non-renewable resources; they are limited and will eventually run out. However, renewable natural resources are things that can grow and be replaced over time if they are carefully managed. Forests, soils and fresh water are renewable.

2.2.2 Global supply

The global distribution of natural resources depends on geology (the materials and rocks that make up the Earth) and climate. Some minerals are rare and are found in only a few locations. For example, **uranium** is found mainly in Australia. Several countries in the Middle East, such as Saudi Arabia and Iran, have rich oil resources but are short of water. Many countries in Africa, such as Botswana, have mineral resources but lack the money to mine and process them.

The human activities of agriculture, fishing, logging and mining all depend directly on natural resources. In developing countries, traditional forms of agriculture such as **subsistence farming** and nomadic herding are still common. These activities are sustainable if farmers move on when an area becomes unproductive, allowing the land to recover. However, poverty and population growth mean that many people now clear forests for farms and overgraze or overcrop small plots of land, resulting in deforestation and land degradation.

Farms in developed countries are usually much larger. For example, the Anna Creek cattle station in South Australia is 24 000 square kilometres, the size of Belgium. In contrast, an average intensive rice farm in Bali is only about one hectare. This is about four times the size of an Australian quarter-acre block of land. Unsustainable agricultural practices in developed countries include the overuse of water, fertilisers and pesticides. For example, fertilisers help crops to grow, but when they end up in rivers and oceans as run-off, they cause algal blooms and damage coral reefs.

FIGURE 3 The process of shifting cultivation means that farmers move on when an area becomes unproductive, allowing the land to recover.

A clearing is made by cutting and burning vegetation. This is known as 'slash and burn'.

Crops are planted and grow well.

After three to four years the nutrients in the soil have been used up and the crops don't grow as successfully.

The clearing is abandoned as farmers move to a new area. As a result the clearing gradually returns to its natural state.

2.2 Activities

To answer questions online and to receive **immediate feedback** and **sample responses** for every question, go to your learnON title at www.jacplus.com.au. *Note:* Question numbers may vary slightly.

Remember

1. What is a natural resource?
2. Outline the difference between renewable and non-renewable resources.
3. List three examples of non-renewable resources that can be recycled.

Explain

4. Which renewable resources are most affected by human activity? Why?
5. When it comes to using natural resources, there are two main problems people face. What are they and why are they important?
6. What does the *sustainable* use of natural resources mean?
7. Why is shifting cultivation a *sustainable* form of agriculture?

Discover

8. Refer to figure 3. Use the internet to research who uses shifting cultivation around the world. Choose one case study and report back to the class about their way of life. Examples of these may include tribes from *places* such as the Amazon, Congo Basin or Papua New Guinea. Compare your chosen tribe's way of life

with your way of life, and explain how it differs when it comes to using resources and accessing food. Upon completion of the presentations, discuss as a class why you think Australian farmers do not use shifting cultivation as their method of agricultural production.

Think

9. What did you have for breakfast today? What resources would have been required to provide it?
10. What are examples of fossil fuels that you use in order to maintain your lifestyle?
11. Create a table that lists 10 renewable and 10 non-renewable resources used by your family. Be specific; for example, list timber used in your furniture. From your list, note some of the waste and pollution that may be created in the use or creation of these resources. How could this be reduced to improve environmental *sustainability*? As a class develop a five-point plan how you could all be more proactive in being more environmentally *sustainable* every day.

2.3 What are Australia's natural resources?

2.3.1 A wide range of resources

Australia has more natural resources per head of population than any other country in the world. The main reason for this is that we have a small population in a very large country.

2.3.2 Minerals

Australia is rich in mineral resources. The Pilbara region in Western Australia, for example, has some of the largest reserves of iron ore in the world. Australia also produces many other minerals, including silver, copper, nickel and tin.

Australia is the world's:
- largest exporter of iron ore, bauxite, lead, diamonds, zinc ores and mineral sands
- second-largest exporter of alumina (processed from bauxite and then turned into aluminium)
- third-largest exporter of gold.

2.3.3 Soils

Australia has generally poor soils, especially when compared with those found in other continents such as North America and Europe. Most Australian soils are low in nutrients, and in some parts of the continent, particularly the more arid areas, high salt content is also a problem. Most parts of Australia are suitable only for sheep and cattle grazing, rather than **intensive agriculture**, owing to low rainfall and poor soils.

There are regions of good soil scattered throughout Australia. These include soils formed from volcanic rock, such as those on the Darling Downs in Queensland and around Orange in New South Wales, and **alluvial soils** (found in river valleys).

2.3.4 Case study: the Pilbara

CASE STUDY

The Pilbara

Most of Australia's iron ore reserves are found in the Pilbara region in north-west Western Australia. The Pilbara accounts for 98 per cent of the country's iron ore production and 96 per cent of its exports. Iron ore is the raw material from which iron is made. Although iron in its cast form has many uses, its main use is in steelmaking. Steel is the main structural metal in engineering, building, ship building, cars and machinery.

Two of the largest companies operating in the Pilbara are BHP Billiton and Rio Tinto. There are many mines in the Pilbara, including those at Mount Tom Price, Marandoo, Channar, Newman and Robe River.

The iron ore in the Pilbara is relatively easy to mine. It is also high quality, so there is strong demand for it from many countries, including Japan, China and South Korea.

FIGURE 1 The Mount Tom Price iron ore mine in the Pilbara

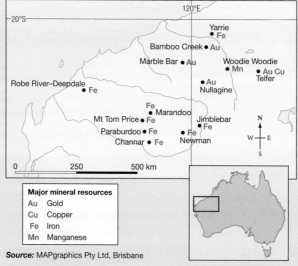

FIGURE 2 Mineral deposits in the Pilbara region

Source: MAPgraphics Pty Ltd, Brisbane

2.3.5 Natural scenery

Australia's spectacular scenery attracts tourists from all over the world, particularly to those sites that are on the World Heritage List. This means that they are recognised as being of global importance due to their great natural or cultural significance.

2.3.6 Forests

Apart from Antarctica, which has no trees, Australia is the world's least forested continent. The most common vegetation in Australia is woodland and shrubland. Before European occupation, about nine per cent of Australia was forested. Today, about five per cent of the country is forested. Even though Australia exports timber products, it also imports a lot of timber, particularly softwoods such as pine.

FIGURE 3 Australia's World Heritage sites

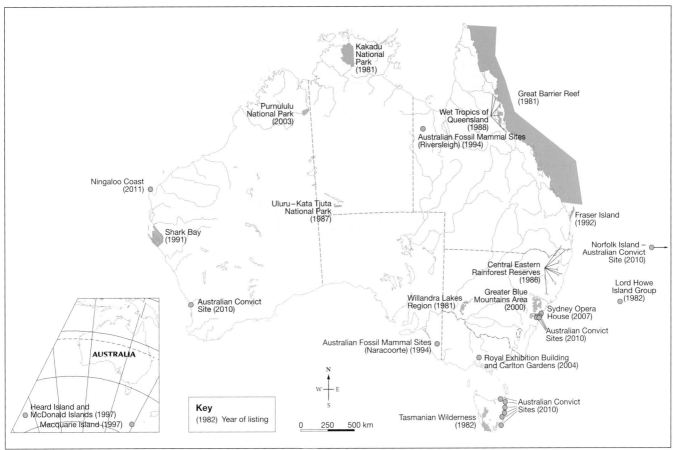

Source: Department of Environment and Water Resources

2.3 Activities

To answer questions online and to receive **immediate feedback** and **sample responses** for every question, go to your learnON title at www.jacplus.com.au. *Note:* Question numbers may vary slightly.

Remember

1. List the main mineral resources produced in Australia.
2. Refer to the case study of the Pilbara.
   (a) Where does Australia's main iron ore production take place?
   (b) What is the main use of iron ore?
   (c) Why is there strong demand overseas for Australian iron ore?
   (d) List three *places* in the Pilbara where there are deposits of iron ore.
3. Refer to figure 3. Which state of Australia has the most World Heritage sites?

Explain

4. Why are many Australian soils suitable only for grazing?
5. Why does Australia's scenery attract many overseas visitors?

Discover

6. Refer to figure 2. If you are at the Mount Tom Price mine, in which direction and how far are the following *places*?
   (a) Yarrie   (b) Robe River–Deepdale   (c) Telfer   (d) Channar?
   What is mined at each location?
7. Using the internet, research the criteria required for an area to become classified as a World Heritage area.

8. In groups of two or three, research one of the World Heritage sites found in Australia. Each group should investigate a different area. Present your group's research back to the class and accompany your presentation with a PowerPoint or other graphic presentation so your peers can see the features that make the *place* so special.

Think

9. What effects does mining have on the natural *environment*? Refer to figure 1. Create a flow diagram that illustrates how these effects relate to each other and how they can have far-reaching consequences beyond the immediate *environment*.

learn on RESOURCES — ONLINE ONLY

🔗 **Explore more with this weblink:** UNESCO criteria

myWorldAtlas — Deepen your understanding of this topic with related case studies and questions.
❯ **World Heritage sites**

2.4 How do we use non-renewable energy?

2.4.1 Australia's energy use

Australia has large reserves of non-renewable energy, such as coal, natural gas, oil and uranium. Over the past 30 years, Australia's energy consumption has increased by over 200 per cent. Most of the energy we use comes from non-renewable sources, particularly coal, which is used for steel manufacturing and 73 per cent of electricity generation.

Each year, over half of Australia's energy products are exported. Australia is the world's fourth largest producer of coal. In the last five years, coal has made up approximately 15 per cent of Australia's exports. Demand for coal from Asia has increased over the last 10 years. Japan is our largest market, claiming 45 per cent of Australian coal. The next biggest importer of Australian coal is China which takes 23 per cent of the market. There is the possibility that demand for Australian coal from south-east Asia will triple in the next 25 years.

FIGURE 1 The biggest buyers in the Australian coal market, 2014

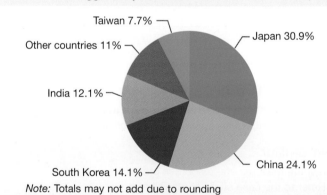

Taiwan 7.7%
Other countries 11%
Japan 30.9%
India 12.1%
China 24.1%
South Korea 14.1%
Note: Totals may not add due to rounding

FIGURE 2 Millions of tonnes of coal are moved around Australia annually.

2.4.2 Global use

The world relies heavily on energy for transport, heating and manufacturing. The amount of energy used varies widely around the world. The world's most commonly used energy sources are oil, coal and gas (all fossil fuels), hydro-electricity and nuclear power.

Internationally, there is more trade in oil than any other product. Oil is unevenly distributed, with most reserves located in the Middle East. As a result, other nations need to buy their oil from this region.

Oil is a fossil fuel and a non-renewable resource. It is believed that oil reserves will eventually run out, probably within 40 to 80 years. Between half and two-thirds of oil production is used in transport. It is also used to produce energy and to **manufacture** products such as plastic, nail polish, lipstick, synthetic textiles and whitegoods.

Every day, 92 million barrels of oil are used around the world; the United States uses 19 million barrels per day while the second highest consumer of oil, China, uses 11 million barrels per day. Australia consumes almost 1 million (or 998 thousand) barrels per day.

Oil is exported to countries that can afford to pay for it. For example, even though the United States has only five per cent of the world's population, it consumes around 20 per cent of global supplies. Some countries import oil because they use more than they produce.

2.4.3 Fossil fuel winners and losers

Governments, oil companies and individuals make billions of dollars from oil. For example, Saudi Arabia's Crown Prince and the Sultan of Brunei are oil billionaires, and large oil companies are some of the most profitable companies in the world.

Oil generates economic growth and can improve people's living standards in producer countries. However, while oil may bring wealth to governments and corporations, wealth does not always trickle down to local populations. Venezuela is one of the world's top 10 oil producers, yet 32 per cent of its population lives below the poverty line.

FIGURE 3 Extraction of coal carries a high environmental cost.

FIGURE 4 Australian non-renewable energy consumption by fuel

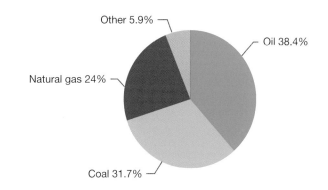

Other 5.9%
Oil 38.4%
Natural gas 24%
Coal 31.7%

FIGURE 5 World energy production

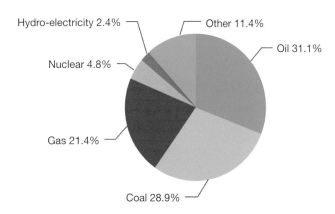

Hydro-electricity 2.4%
Other 11.4%
Nuclear 4.8%
Oil 31.1%
Gas 21.4%
Coal 28.9%

FIGURE 6 Global oil consumption

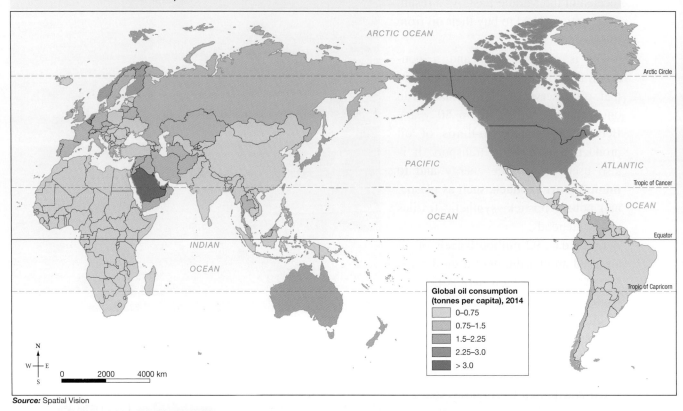

Global oil consumption
(tonnes per capita), 2014
	0–0.75
	0.75–1.5
	1.5–2.25
	2.25–3.0
	> 3.0

Source: Spatial Vision

2.4 Activities

To answer questions online and to receive **immediate feedback** and **sample responses** for every question, go to your learnON title at www.jacplus.com.au. *Note:* Question numbers may vary slightly.

Remember

1. Refer to figure 1.
   (a) In which geographic region are these nations located?
   (b) How many more per cent of coal does China import than Taiwan?
2. Other than energy, what other products are manufactured from oil?
3. How many more barrels of oil does the United States use than Australia per day?

Explain

4. Why doesn't the money made from the sale of fossil fuels, such as oil, always benefit the local people in the country that sells the product?

Discover

5. The **extraction** of fossil fuels can have negative impacts on the *environment*. Using the internet, research the 2010 BP oil disaster off the coast of Mexico. Create a poster that illustrates the causes of the event and its effects, and explains the clean-up methods used to try to contain the damage. Ensure your presentation is visually appealing by using a variety of maps, photos and graphs.

Predict

6. What will happen to Australian coal reserves if demand from other countries continues to grow? What could be some of the impacts of this? Does this raise any ethical dilemmas?

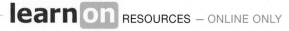 RESOURCES — ONLINE ONLY

  Explore more with these weblinks: Gulf of Mexico, NOAA, Treehugger

2.5 Should Australia focus on using renewable energy?

2.5.1 Australia's renewable energy

Australia has many potential renewable energy sources: hydro-electric, wind, solar, tidal, geothermal and **biomass** energy. Although the use of these is increasing, renewables provide us with only about six per cent of our energy needs. In 2015, the Australian Government reviewed its Renewable Energy Target and set the goal that by 2020 at least 33 000 gigawatt-hours, or 23.5 per cent of Australia's electricity, would come from renewable sources. This would be enough electricity to power approximately five million homes for a year. Other benefits may include billions of dollars of investment, the creation over 15 000 jobs and, in the long term, it could even save every Australian household $140 off their electricity bill annually.

Australia is a large country with a number of different environments that are suited to supplying renewable energy. There has been an increase in public and government concern over global warming and greenhouse gas emissions, so more research into and development of renewable energy resources may help to minimise these concerns.

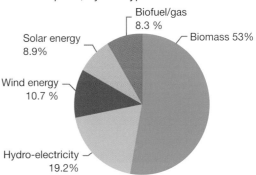

FIGURE 1 Australia's renewable energy consumption, by fuel type

- Biofuel/gas 8.3 %
- Biomass 53%
- Solar energy 8.9%
- Wind energy 10.7 %
- Hydro-electricity 19.2%

Note: Totals may not add due to rounding.

2.5.2 Different types of energy

Some types of renewable energy that could be sustainable in Australia include the following.

- *Geothermal energy.* Australia has huge underground energy resources known as 'hot rocks'. Water can be heated by pumping it underground through these hot rocks. The resulting steam then drives a **turbine** to generate electricity.
- *Wind power.* A typical wind turbine can meet the energy needs of up to 1000 homes.
- *Tidal power.* Waves can drive turbines to produce electricity. Tidal power is especially suitable for powering **desalination** plants, though Australia does not use this system.
- *Biomass energy.* This is produced from the combustion of organic matter, such as sugar cane and corn crops. Biomass can be used to produce electricity as well as liquid fuels like ethanol and biodiesel.
- *Solar power.* Solar energy technologies harness the sun's heat and light to provide heating, lighting and electricity. Two types of solar technologies are currently under development in Australia.
  - Photovoltaic cells convert solar energy directly into electricity. These can be placed on roofs in order to collect direct sunlight.
  - Solar thermal systems use the sun's heat to generate electricity by first heating a fluid such as water to create steam, which drives a turbine to generate electricity. Australia has abundant solar radiation, and therefore great potential for the development of solar energy. Germany is currently the highest consumer of solar electricity in the world.
- *Hydro-electric power.* Most of Australia's hydro-electric power is generated in Tasmania and by the Snowy Mountains Hydro-Electric Scheme in New South Wales. Around six per cent of Australia's electricity comes from hydro-electric power, but there are limited opportunities for increasing this because of the lack of water resources, and because building dams can be controversial.

The use of sustainable energy is still in its early stages but is growing rapidly in China, the United States and Europe. Fossil fuels are currently cheaper and more convenient to produce than renewable energy sources. The reason for this is that we do not pay the real cost of their use — we do not pay for the huge cost of releasing waste products into the **atmosphere**. In future, a carbon tax or restrictions on the use of fossil fuels will increase their cost, perhaps making renewable energy a more attractive option for consumers.

FIGURE 2 Some sources of renewable energy: (a) solar, (b) biomass, (c) wind, (d) hydro-electric (e) geothermal, (f) tidal

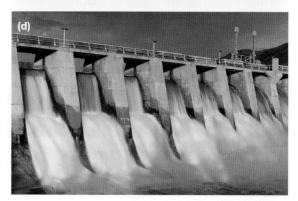

2.5 Activities

To answer questions online and to receive **immediate feedback** and **sample responses** for every question, go to your learnON title at www.jacplus.com.au. *Note:* Question numbers may vary slightly.

Remember

1. In which renewable energy resources is Australia (a) rich and (b) poor?
2. Define the term 'renewable energy'. List the main forms of renewable energy.
3. What are the two types of solar technology?

Explain

4. Why is renewable energy not widely used in many countries?

Discover

5. Refer to the pie graph in figure 1.
   (a) What type of renewable resource provides the greatest source of energy in Australia?
   (b) Australia is a sunny country, so why does only 8.9 per cent of our country's renewable energy come from the sun?

Predict

6. List some of the likely positive and negative consequences if Australia stopped using non-renewable energy resources altogether and replaced them with renewable energy sources.

 my**World**Atlas   Deepen your understanding of this topic with related case studies and questions.
❂ **Energy in Australia**

2.6 SkillBuilder: Constructing a pie graph

WHAT IS A PIE GRAPH?

A pie chart, or pie graph, is a graph in which slices or segments represent the size of different parts that make up the whole. The size of the segments is easily seen and can be compared. Pie graphs give us an overall impression of data.

Go online to access:

- a clear step-by-step explanation to help you master the skill
- a model of what you are aiming for
- a checklist of key aspects of the skill
- a series of questions to help you apply the skill and to check your understanding.

FIGURE 1 Percentage of electricity generated from renewables in Australia by energy source (2010)

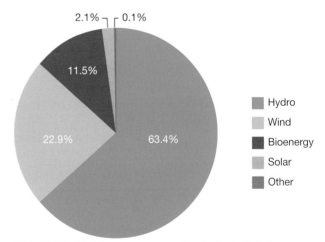

- Hydro
- Wind
- Bioenergy
- Solar
- Other

Note: 5.2% of total energy consumption in Australia is from renewable sources.

Source: Clean Energy Council, Clean Energy Australia 2010, page 6

 **learn** on RESOURCES — ONLINE ONLY

Watch this eLesson: Constructing a pie graph (eles-1632)

Try out this interactivity: Constructing a pie graph (int-3128)

2.7 Is water a renewable resource?

2.7.1 Water as a resource

The amount of water on Earth has not changed since the beginning of time; there is only a finite, or fixed, amount. The water used by ancient and extinct animals and plants millions of years ago is the same water that today falls as rain. The amount of water cannot be increased or decreased. It is cycled and recycled, and constantly changes its state from gas, to liquid, to solid, and back.

Water is a resource. Like any other resource, it has no value in itself, but has great value when a use is found for it. Environments where water is found are also a resource. A river can be the site of a settlement that provides transport as well as food. A **riverine environment** that includes fish, birds, wildlife, wetlands, plants and micro-organisms is also valuable as a living system and can therefore be regarded as a resource.

2.7.2 The water cycle

All the water on Earth moves through a cycle that is powered by the sun. This cycle is called the water cycle, or **hydrologic cycle**. Water is constantly changing its location (through constant movement) and its form (from gas, to liquid, to solid). Evaporation, condensation and freezing of water occur during the cycle.

2.7.3 How long does water stay in the one place?

Water can stay in the one place very briefly or it can stay for many thousands of years. It has been calculated that water stays in the atmosphere for an average of nine days before it falls again to Earth as precipitation. Water stays in soil for between one and two months. If you live in an area that has experienced drought or a very long summer without rain, you may have noticed that the soil dries out and forms cracks. Once the seasons change and it begins to rain, the soil absorbs water again and the cracks disappear.

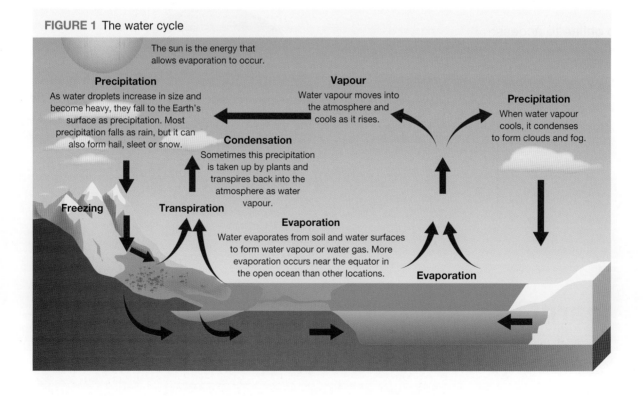

FIGURE 1 The water cycle

The sun is the energy that allows evaporation to occur.

Precipitation
As water droplets increase in size and become heavy, they fall to the Earth's surface as precipitation. Most precipitation falls as rain, but it can also form hail, sleet or snow.

Vapour
Water vapour moves into the atmosphere and cools as it rises.

Precipitation
When water vapour cools, it condenses to form clouds and fog.

Condensation
Sometimes this precipitation is taken up by plants and transpires back into the atmosphere as water vapour.

Freezing

Transpiration

Evaporation
Water evaporates from soil and water surfaces to form water vapour or water gas. More evaporation occurs near the equator in the open ocean than other locations.

Evaporation

Water spends between two and six months in snow and rivers but a lot longer in large lakes, glaciers, oceans and **groundwater**. The longest time water stays in one place is in the Antarctic ice sheets. Some ice core samples in Antarctica contain water that is 800 000 years old, but the average is about 20 000 years.

The length of time water spends as groundwater can be an average of 10 000 years if it is very deep, but it can stay much longer.

FIGURE 2 A scientist working with ice core samples in Antarctica. Some of the longest records of our climate have come from large ice sheets over three kilometres thick in Greenland and Antarctica. They produce records going back several hundred thousand years.

2.7 Activities

To answer questions online and to receive **immediate feedback** and **sample responses** for every question, go to your learnON title at www.jacplus.com.au. *Note:* Question numbers may vary slightly.

Remember

1. Is water a renewable or a non-renewable resource?
2. List all the ways that water can be used as a resource.
3. Name two *places* where water stays in the same *place* for the longest.

Explain

4. Using the information about the water cycle, write and present 'The incredible journey of water'. Focus on various *interconnections* between water and the *environment*. Include diagrams and photographs, and present your story electronically, as a poster, a written piece, a drama piece or a song.
5. How does the water cycle prove that we are using the same water that the dinosaurs used — in other words, that it is finite (limited)?

Discover

6. Use the internet to find all the information that is discovered by reading ice core samples. Choose two of the most interesting facts and share them with your class on a blog or wiki.

Predict

7. With a global population that is increasing by about 75 million people each year, how will it be possible for the finite water on Earth to be shared fairly?

 RESOURCES — ONLINE ONLY

✦ **Try out this interactivity:** Water works (int-3077)

2.8 SkillBuilder: Annotating a photograph

USING ANNOTATED PHOTOGRAPHS IN GEOGRAPHY

Photographs are used to show aspects of a place. **Annotations** are added to photographs to draw the reader's attention to what can be seen and deduced.

Go online to access:

- a clear step-by-step explanation to help you master the skill
- a model of what you are aiming for
- a checklist of key aspects of the skill
- a series of questions to help you apply the skill and to check your understanding.

FIGURE 1 Campaspe River near Axedale

Power line

Riparian vegetation

Sandy bank

Snag

Water pump input

River

Source: Taken by Mattinbgn, 17 March 2012. © Creative Commons

learn on RESOURCES — ONLINE ONLY

Watch this eLesson: Annotating a photograph (eles-1633)

Try out this interactivity: Annotating a photograph (int-3129)

2.9 How is groundwater used as a resource?

2.9.1 What is groundwater?

An important part of the water cycle, groundwater is the water that is found under the Earth's surface. Many settlements — especially those in arid and semi-arid areas — rely on groundwater for their water supply.

When rain falls to the ground, some flows over the surface into waterways and some seeps into the ground. Any seeping water moves down through soil and rocks that are permeable; that is, they have pores that allow water to pass through them. Imagine pouring water into a jar of sand or pebbles; the water would settle into the spaces between the sand or stones.

Groundwater is water held within water-bearing rocks, or **aquifers**, in the ground. These work like sponges. They hold water in the tiny holes between the rock particles.

2.9.2 Artesian water

An **artesian aquifer** occurs between impermeable rocks, and this creates great pressure. When a well is bored into an artesian aquifer, water often gushes out onto the surface. This flow will not stop unless the water pressure is reduced or the bore is capped (sealed).

Groundwater and surface water are interconnected — they depend on each other. Groundwater is only replenished when surface water seeps into aquifers. This is called groundwater recharge, and it is affected by whether there is a lot of rain or a drought is occurring.

FIGURE 1 An artesian aquifer

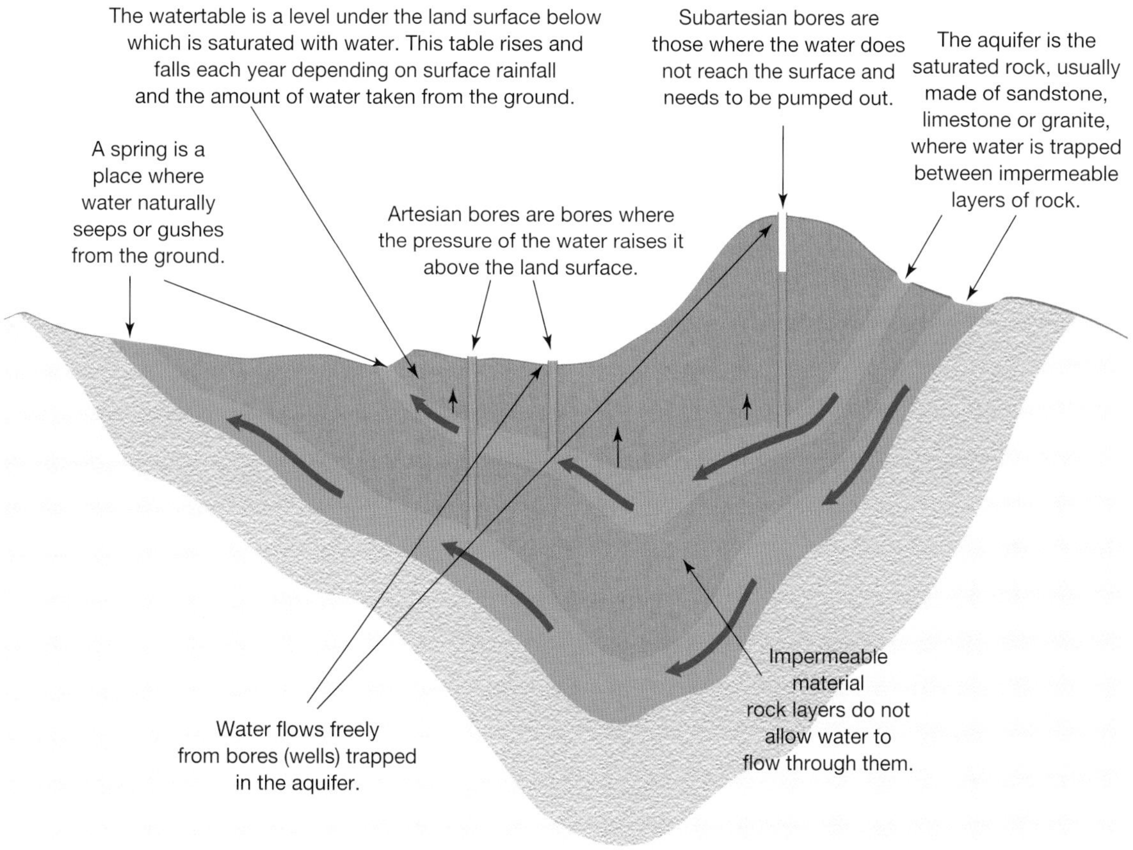

The watertable is a level under the land surface below which is saturated with water. This table rises and falls each year depending on surface rainfall and the amount of water taken from the ground.

Subartesian bores are those where the water does not reach the surface and needs to be pumped out.

The aquifer is the saturated rock, usually made of sandstone, limestone or granite, where water is trapped between impermeable layers of rock.

A spring is a place where water naturally seeps or gushes from the ground.

Artesian bores are bores where the pressure of the water raises it above the land surface.

Water flows freely from bores (wells) trapped in the aquifer.

Impermeable material rock layers do not allow water to flow through them.

FIGURE 2 The water in this mound spring in South Australia has taken over two million years to move to the surface from recharge areas innorthern Queensland. It can take up to 1000 years to move about one metre.

2.9.3 Global groundwater resources

Approximately two per cent of the Earth's water occurs as groundwater, compared with 0.1 per cent as rivers and lakes and 94 per cent as oceans.

About 1.5 billion people in the world rely on groundwater for their survival. Some groundwater is fresh and can be used for drinking. Other groundwater can be brackish or even saltier than the sea.

FIGURE 3 The world's major groundwater basins

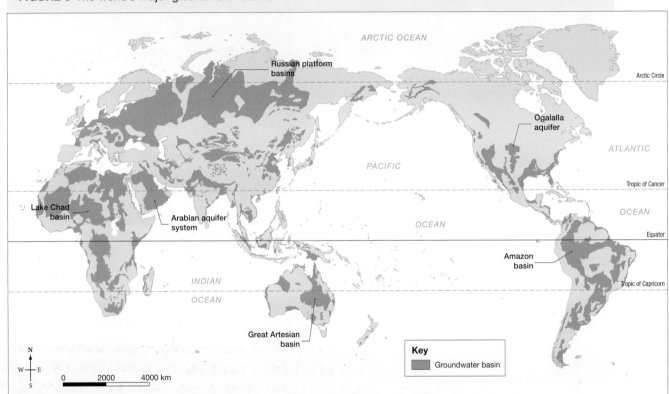

Source: BGR & UNESCO 2008: Groundwater Resources of the World 1 : 25 000 000. Hannover, Paris.

Groundwater is vital for drinking, irrigation and industry use. Some industries bottle and sell spring and mineral water, and make soft drinks and beer. Bore water is used to water suburban gardens and parks, golf courses and crops. Groundwater is also important to the natural environment in wetlands and in supporting unique plants and animals. Groundwater keeps many of our rivers flowing, even when there are long periods without rain.

Troubled waters

For many years now, more and more water has been taken out of the ground. People believed it was unlimited, but it is in danger of running out in some areas, owing to the large number of wells pumping water.

If people use more groundwater than is being recharged, aquifers may dry up. Groundwater is very slow-moving and can take many years to move into deep aquifers. For this reason, groundwater is a finite and non-renewable resource, and is often referred to as *fossil water.*

2.9 Activities

To answer questions online and to receive **immediate feedback** and **sample responses** for every question, go to your learnON title at www.jacplus.com.au. *Note:* Question numbers may vary slightly.

Remember

1. What does the word permeable mean?
2. What is groundwater recharge?

Explain

3. What is the difference between an aquifer and an artesian aquifer? Use a diagram to help you.
4. Draw a diagram to show how surface water reaches the watertable to become groundwater.
5. Describe conditions that might result in a watertable rising or falling.
6. Outline how groundwater and surface water are *interconnected*.
7. Describe the groundwater resources in North Africa and West Asia (the Middle East). (Use the **Regions** resource in the Resources tab.)

Discover

8. Refer to figure 3 to describe the location of the world's groundwater regions.

Predict

9. Imagine that the world's groundwater continues to be used faster than it can be replaced. Use the **Bubbl.us** weblink in the Resources tab to brainstorm all the possible consequences this will have on people and the *environment*.

Think

10. Water is a renewable resource. Why is groundwater sometimes thought of as fossil water and as a non-renewable resource? Write your answer as a newspaper article titled 'Out of sight, out of mind'.
11. Discuss in small groups the fairness of taking too much water from old aquifers. Should this type of water use be restricted? Who should make this decision?

my **World** Atlas — Deepen your understanding of this topic with related case studies and questions.
❯ **Salisbury Council — Aquifer storage, transfer and recovery**

2.10 How do Aboriginal peoples use groundwater?

2.10.1 Groundwater sources

Indigenous Australian peoples have lived in the Australian landscape since the beginning of the Dreaming, thousands of years by European estimates, and they have had the knowledge to survive many changes and challenges. In order to obtain water in the country's dry regions, particularly in Australia's deserts, they have needed to know where to find groundwater.

There are many groundwater sources throughout Australia that have long been used by Indigenous Australian nations. One of these sources is **soaks**: groundwater that comes to the surface, often near rivers and dry creek beds, and which can be identified by certain types of vegetation. Another source is **mound springs**: mounds of built-up minerals and sediments brought up by water discharging from an aquifer.

Mound springs of the Oodnadatta Track

The Oodnadatta Track is located in the north-east of South Australia. The track follows the edge of the Great Artesian Basin and the south-western edge of Kati Thanda–Lake Eyre and, along its route, groundwater makes its way to the surface in several locations.

The Oodnadatta Track crosses the traditional lands of three Aboriginal nations. In the south, between Lake Torrens and Kati Thanda–Lake Eyre, are the Kuyani people; most of the west of Kati Thanda–Lake Eyre is the land of the the Arabana people; and to the north is the land of the Arrernte people.

Many springs have cultural significance today for local Aboriginal peoples, whose ancestors relied on the springs as water sources and as sacred sites for important ceremonies. Knowledge of the springs in this region has been passed down over many generations through **Dreaming** Stories.

This knowledge was also passed on to explorers and colonisers. John McDouall Stuart followed this track to complete the first crossing of Australia's interior from south to north in 1862; the overland telegraph was constructed along its pathway; and the Great Northern Railway, which made the land of the Northern Territory accessible for European occupation, followed the same route.

Mound springs were very important for Indigenous Australian peoples. They could rely upon springs as reliable sources of water in a very harsh, dry environment.

FIGURE 1 Location of the Oodnadatta Track and Great Artesian Basin, one of the world's largest groundwater basins

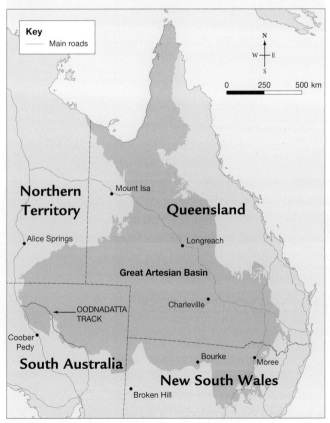

Source: Spatial Vision / Geoscience Australia

FIGURE 2 The Oodnadatta Track passes close to the Old Ghan, the Great Northern Railway.

Old campsites and animal remains provide evidence that they remained there for varying time periods. However, because the plant and animal life around these regions is quite sparse, people had to move regularly and travel away from the springs when rainfall allowed that to occur.

Because the springs were strung out over hundreds of kilometres, they were also part of an important network of trading and communication routes across Australia. As Aboriginal peoples moved around the region, they traded goods and communicated with other Aboriginal nations. This interconnection allowed them to trade resources such as ochre, stone and wooden tools, bailer shells and pituri. Pituri is a spindly shrub used by Indigenous Australian peoples during ceremonies and to spike waterholes to catch animals for food.

2.10.2 Case study: Locating water

CASE STUDY

Stories help map the location of water

Indigenous Australian knowledge of the land and how to survive in it has been passed from generation to generation through Dreaming Stories. During the dry seasons and periods of drought, Aboriginal peoples congregated at the mound springs. These springs were linked by Aboriginal songs and Dreaming Stories, and are often connected to rain-making rituals.

FIGURE 3 Groundwater springs along the Oodnadatta Track

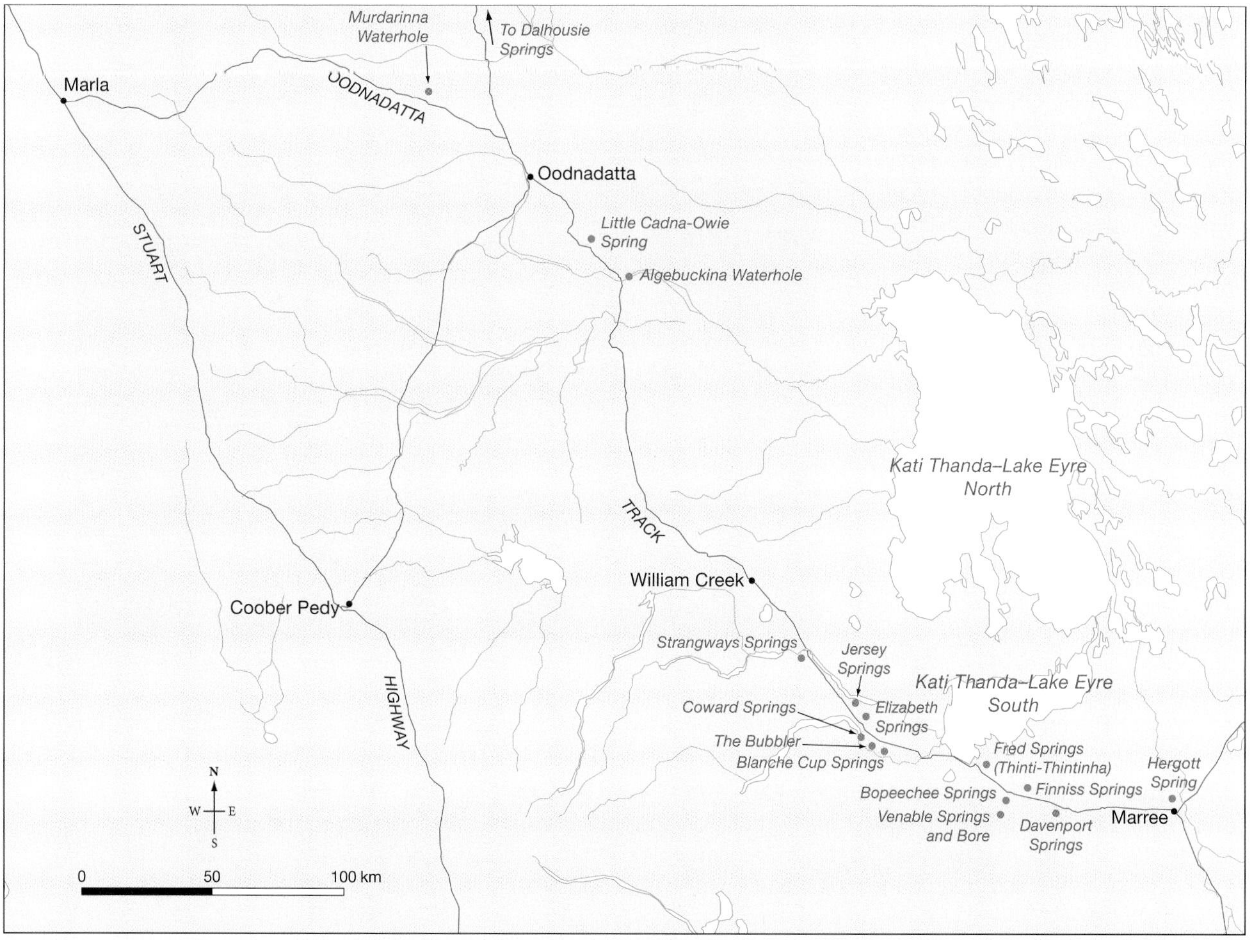

Source: Redrawn with permission from the SA Arid Lands Natural Resources Management Board / © Copyright Commonwealth of Australia Geoscience Australia 2006

Dreaming Stories

1 Thutirla Pula (Two Boys Dreaming)

This is one of the most important stories of the Wangkangurru and other people of Central Australia. Thutirla Pula is how the spirits of the Dreaming first crossed the desert they call Munga-Thirri (land of sandhills). The story tells of two boys crossing the Simpson Desert, through Queensland and back to just north of Witjira (Dalhousie) in the Finke River area. The songline contains information on every waterhole or soak that was known in the Simpson Desert. Following this songline meant you could cross the Simpson Desert using available groundwater along the way, taking 600 kilometres off the usual journey south of the Simpson Desert to Kati Thanda–Lake Eyre, then back north along the Diamantina River.

2 Bidalinha (or the Bubbler)

The Kuyani ancestor Kakakutanha followed the trail of the rainbow serpent Kanmari to Bidalinha (or the Bubbler) where he killed it. He then threw away the snake's head, which is represented by Hamilton Hill, and cooked the body in a dirga, or oven, which is now Blanche Cup. Kakakutanha's wife, angry at missing out on the best meat from the snake, cursed her husband, and he went on to meet a gruesome death at Kudna-ngampa (Curdimurka). The bubbling water represents the movements of the dying serpent.

3 Thinti-Thintinha Spring (Fred Springs)

The willy wagtail (or thunti-thuntinha) danced his circular dance to create this spring and the surrounding soils, which are easily airborne in windy conditions. The moral to the story is that while it is easy to catch the skilful little willy wagtail, you must never do so because of the terrible dust storms that may follow.

4 Kewson Hill: the Camp of the Mankarra-kari — the Seven Sisters

The Seven Sisters came down here to dig for bush onions (yalkapakanha). As they peeled the onions, they tossed the skins to one side, creating the dark-coloured extinct mound spring on the south-west side of the track. The peeled bulbs created the light-coloured hill (yalka-parlumarna) to the north-east, also an extinct mound spring.

5 Dalhousie Springs

Dalhousie Springs is a popular oasis in the arid desert region of the northernmost part of South Australia.

FIGURE 4 The Old Bubbler on the Oodnadatta Track

FIGURE 5 Fed by the thermal waters of the Great Artesian Basin, the water in Dalhousie Springs is between 34 and 38°C.

FIGURE 6 Cross-section of a typical mound spring

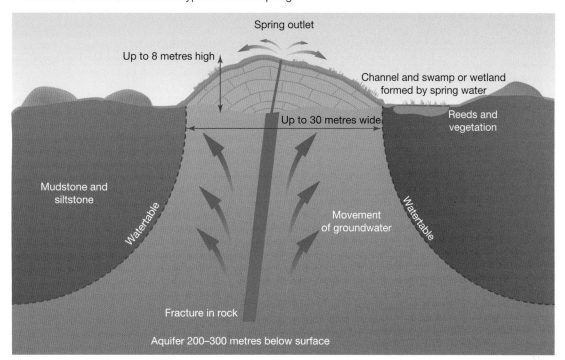

2.10 Activities

To answer questions online and to receive **immediate feedback** and **sample responses** for every question, go to your learnON title at www.jacplus.com.au. *Note:* Question numbers may vary slightly.

Remember

1. Why were Dreaming Stories important to Aboriginal peoples in the Oodnadatta region?

Explain

2. Why is groundwater so important to the communities in this region today? Describe the *interconnections* between groundwater and people.
3. How do Dreaming Stories help map the groundwater in this region?

Discover

4. Use Google Earth and enter the search terms Oodnadatta or William Creek to locate the Oodnadatta Track. Describe the landscape you see. Why is finding groundwater so important in this *environment*?
5. Use an atlas or Google Earth to locate Dalhousie Springs, Birdsville, Kati Thanda–Lake Eyre and the Diamantina River. See if you can follow the Two Boys Dreaming Story and show how following the story saved a lot of time travelling across the desert.
6. How do Dreaming Stories help identify the cultural value placed on these water *environments*?

 RESOURCES — ONLINE ONLY

 Try out this interactivity: Oodnadatta Track (int-3079)

2.11 Review

2.11.1 Review
The Review section contains a range of different questions and activities to help you revise and recall what you have learned, especially prior to a topic test.

2.11.2 Reflect
The Reflect section provides you with an opportunity to apply and extend your learning.

Access this subtopic at **www.jacplus.com.au**

TOPIC 3
Our blue planet: water

3.1 Overview

Numerous **videos** and **interactivities** are embedded just where you need them, at the point of learning, in your learnON title at www.jacplus.com.au. They will help you to learn the content and concepts covered in this topic.

3.1.1 Introduction

Viewed from space, the Earth is a sphere of blue. Water covers most of our planet. We depend on water for life; in fact, no life is possible without it. Water is a precious and finite resource, yet most of the Earth's water is too salty for humans, animals or plants to use. The amount of available fresh water on Earth needs to be shared among an ever-growing global population. Access to water is a basic human right. It is a resource that must be used carefully so that current and future populations can have adequate supplies.

Starter questions

1. List the main ways that you use water.
2. List two natural **environments** that have a lot of water. Name where they are or show their location on a map.
3. List two natural environments that have very little water. Name where they are or show their location on a map.
4. Why do you think water is thought of as a precious resource? Justify your answer.
5. List all the ways that water is used by people, animals, plants and the environment.
6. How much of the Earth is covered by water? How much of this water can be used by people?

This picture shows what all of Earth's water would look like if it was contained in a sphere, in comparison with the size of the Earth. The blue sphere representing all of Earth's water has a diameter of 1385 kilometres.

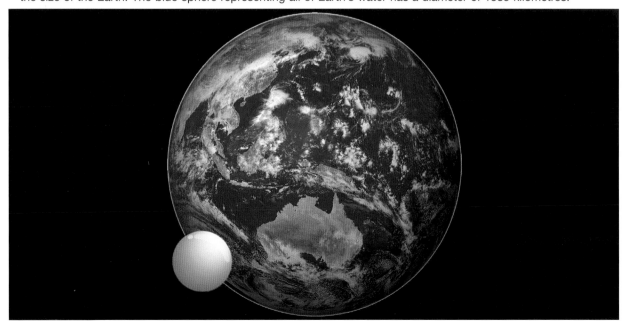

learn on RESOURCES — ONLINE ONLY

🔗 **Explore more with this weblink:** Introduction to water

3.2 How much water is there in the world?

3.2.1 The world's water

Water is vital to our survival and essential to most human activities. Although Earth seems blue in space, not much of the water we see is available for use. And of the useable fresh water that can be seen, access to it is unequal across the globe.

Water covers about 75 per cent of the Earth's surface. Yet, as figure 1 shows, almost all this water (97.5 per cent) is salt water and not available for human consumption. Only 2.5 per cent of the world's water is fresh, but most of this is also unavailable for use by people. More than two-thirds (69.5 per cent) of this fresh water is locked up in glaciers, snow, ice and permafrost. Of the remaining amount, 30.1 per cent is found in groundwater. Only 0.4 per cent is left — found in rivers, lakes, wetlands and soil as well as in the bodies of animals and plants.

FIGURE 1 The distribution of water on Earth

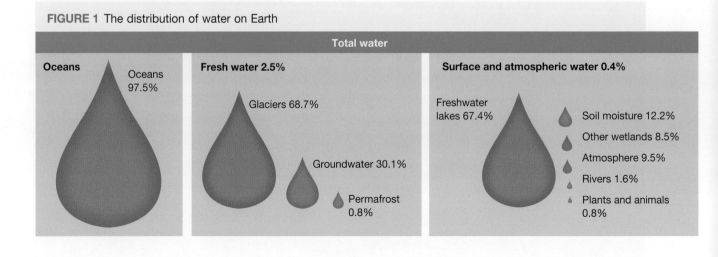

FIGURE 2 The distribution of global rainfall

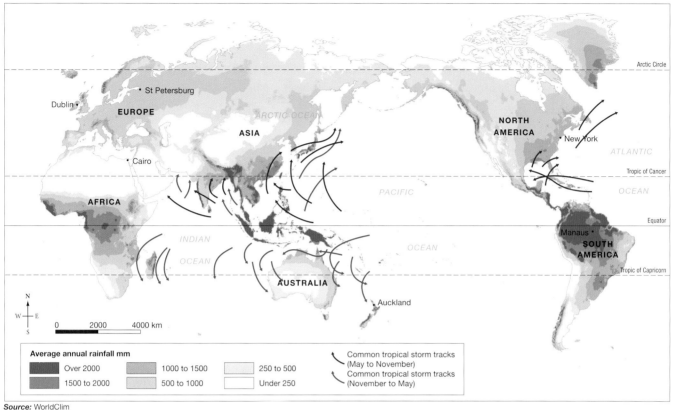

Source: WorldClim

Global rainfall

The Earth's water is constantly moving. Rainfall patterns show which world regions receive more rain than others. The amount of rainfall, or **precipitation**, is related to the amount of water available for use by people.

Green and blue water

The key to our survival is being able to use the water that falls on land and into rivers and streams. Water is sometimes categorised as either **blue water** or **green water**.

Green water is the water that does not run into streams or recharge groundwater but is stored in the soil or stays on top of the soil or vegetation. This water eventually evaporates or transpires through plants. Green water is used by crops, forests, grasslands and savannas.

The amount of blue and green water available changes throughout the year, from year to year, and according to changes in the environment.

3.2.2 Climate change and impact on rainfall and run-off

The majority of climate scientists believe that **climate change** will have an impact on rainfall patterns and **run-off**. Climate models have shown that areas in the northern latitudes are likely to experience more rain, and areas closer to the equator and mid-latitudes will receive less rain. Some regions will experience droughts, while others will experience high rainfall and even flooding.

Already, in the last 100 years, global rainfall patterns have changed. In some areas such as North America, South America, northern Europe, and northern and central Asia, rainfall has increased significantly. In other areas such as the Sahel, the Mediterranean, southern Africa, and parts of Asia, rainfall has decreased.

FIGURE 3 Predicted change in annual run-off due to climate change, 2084

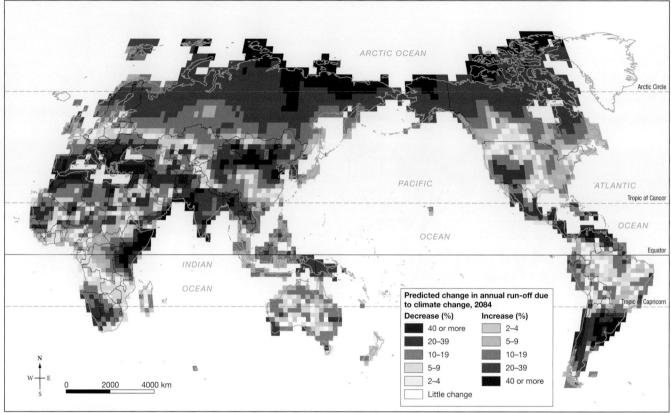

Predicted change in annual run-off due to climate change, 2084

Decrease (%)
- 40 or more
- 20–39
- 10–19
- 5–9
- 2–4
- Little change

Increase (%)
- 2–4
- 5–9
- 10–19
- 20–39
- 40 or more

Source: Geophysical Fluid Dynamics Labratory, National Oceanic and Atmospheric Administration

3.2 Activities

To answer questions online and to receive **immediate feedback** and **sample responses** for every question, go to your learnON title at www.jacplus.com.au. *Note:* Question numbers may vary slightly.

Remember

1. What percentage of the world's water is:
   (a) salty   (b) available for use by people?
2. How will climate change affect rainfall patterns?

Explain

3. Study figures 1 and 3 and use the **Regions** resource in the Resources tab.
   (a) Describe how much rain falls in North Africa and West Asia (the Middle East). How does this compare with Australia?
   (b) What is predicted to happen to annual run-off in these regions as a result of climate change? What impact might this have on people and the *environment*?
4. Write a statement about the *interconnection* between high rainfall and location at the equator and mid-latitudes. Name two *places* that do not fit this pattern.
5. Use the text to outline the difference between blue and green water. List two things that might change the amount of blue and green water available.
6. If farmers use irrigation, what type of water would they rely on? What about farmers who do not have access to irrigation?

Predict

7. Work in groups of three to list what might happen to people and the *environment* in regions that:
   (a) will receive more rainfall than they do now
   (b) will receive less rainfall than they do now.
   Complete a consequence chart for each change.

Think

8. Study figure 3 and an atlas.
   (a) Name three *places* that are predicted to receive more run-off due to climate change.
   (b) Name three *places* that are predicted to receive less run-off due to climate change.
   (c) Compare these six *places* with the global rainfall map, figure 1. Which of the following statements is true?
   - Most *places* with very low rainfall have lower run-off.
   - All *places* with very high rainfall experience increased run-off.
   - The *places* with the greatest *change* in run-off will be northern Russia and northern Canada.

   Rewrite any false statements and make them true.

3.3 What are some amazing water facts?

3.3.1 Wonderful water

Although only 0.4 per cent of Earth's water is available for use by people, plants and animals, this amounts to over 117 trillion litres — on a scale too big for us to really appreciate. There are some other amazing water facts that show how much water there is, how water connects places and how water is used. Use the **Planet Earth, fresh water** weblink in the Resources tab to find out more about water.

The world's largest swimming pool

The artificial lagoon in Citystars Resort, Sharm el-Sheikh, Egypt, covers an area of 9.68 hectares. It is a full kilometre longer than the previous record holder in San Alfonso del Mar resort, Algarrobo, Chile, which measures 1013 metres in length and holds 250 000 cubic metres of water (see figure 1).

The world's largest freshwater river

The Amazon River in South America is the second longest river in the world and is by far the largest by water flow. It has the largest **drainage basin** in the world and approximately one-fifth of the world's total river flow.

FIGURE 1 Gargantuan pools are on the rise in luxury resorts around the world.

FIGURE 2 The river with the world's greatest water flow is the Amazon River.

The **discharge** from it is about 7000 cubic kilometres every year, or 219 million litres of water every second. In contrast, Australia's largest river in terms of water flow, the Mitchell River, discharges an average of only 12 cubic kilometres each year.

The waterfall with the most water

The flow rate at Niagara Falls' main falls has been measured at around 2.6 million litres per second — enough water to fill an Olympic swimming pool in about one second.

FIGURE 3 Niagara Falls

The place on Earth with the most rainy days

The wettest place in the world (based on the average number of rainy days received each year) is Mount Wai'ale'ale in Hawaii. The summit is 1569 metres above sea level and receives over 350 days of rain each year. Mount Wai'ale'ale records an average of up to 13 000 millimetres of rain per year. In some years, rain has been known to fall for 360 days per year!

FIGURE 4 Mount Wai'ale'ale averages over 350 rainy days each year.

The wettest place in the world

The wettest place in the world (based on the yearly average total) is Mawsynram, India, which receives an average of 11 870 millimetres (nearly 12 metres) of rain each year. It has a subtropical highland climate with a long monsoon season.

How much water is there in the atmosphere?

If it was possible to rain across the whole planet at one time, there is enough water in the atmosphere at any time to produce about 2.5 centimetres (25 millimetres) of rain over the whole surface of the Earth.

FIGURE 5 Women in the potato fields of Mawsynram wear rain protection made of leaves and cane.

FIGURE 6 A lot of water is held in the atmosphere.

The biggest dam

The Three Gorges Dam in China is the world's largest dam and hydro-electricity plant.

FIGURE 7 The Three Gorges Dam produces the most hydro-electricity in the world.

Where is the highest number of desalination plants in the world?

Places with little water often build desalination plants to convert sea water to fresh water. The Persian Gulf has the highest number of desalination plants in the world, with Saudi Arabia the world's largest producer.

FIGURE 8 Al-Jubail, the largest desalination plant in the world, is located on the Persian Gulf in Saudi Arabia.

3.3 Activities

To answer questions online and to receive **immediate feedback** and **sample responses** for every question, go to your learnON title at www.jacplus.com.au. *Note:* Question numbers may vary slightly.

Remember

1. Watch the first half of the video clip on the Iguazu Falls (**Iguazu Falls** weblink in the Resources tab).
   (a) Where are the Iguazu Falls located?
   (b) How wide are these falls?
   (c) How much water flows over the falls each second when they are in flood?
   (d) What is the size or *scale* of the flooded area of the Parana River?

Explain

2. How many Olympic-sized swimming pools would fit into the world's largest swimming pool? Find its location on Google Earth or in an atlas.

Discover

3. Conduct some research to find the location and name of the longest river in the world and the river with the second-highest discharge.
4. How much does it rain where you live? How does this compare with the wettest place in the world?
5. Use a blank outline world map to locate the *places* mentioned in this section. Annotate your map with the water facts for which each *place* is famous.
6. Climate maps give you clues about why each location holds amazing records. Look at your atlas and match the location of the *places* with these facts. Make a list of the geographic features that contribute to each amazing fact.
7. Conduct some research to find the following information: the *place* in the world where the most rain fell in a single 24-hour period; the *place* in Australia that has the record for the wettest year; the wettest town in Australia. Compare each of these with where you live.

 **learn**on RESOURCES — ONLINE ONLY

🔗 **Explore more with this weblink:** Planet Earth, fresh water

🔗 **Explore more with this weblink:** Iguazu Falls

3.4 How does Australia's climate affect water availability?

3.4.1 Dry, variable and evaporated

Australia is the driest inhabited continent (only Antarctica is drier), and there is very little fresh water available for our use. Rain falls unevenly across the country and from season to season.

The driest part of Australia is around the Lake Eyre Basin, and the wettest locations are places in north-east Queensland and western Tasmania. Rainfall is also highly variable, meaning that the amount of rainfall can vary or change from year to year. For example, one year a location might have very high rainfall, the next year it might have very low rainfall, and the following might be an average year.

3.4.2 Variability

Rainfall variability is the way rainfall totals in a given area vary from year to year. For example, if an area has low rainfall variability, it means rainfall will tend to be fairly consistent from one year to the next. Many coastal areas show this kind of rainfall pattern. In contrast, high rainfall variability means rainfall is likely to be irregular from one year to the next; there may be heavy rainfall in some years and little or no rainfall in others. Desert areas in central Australia tend to have low rainfall and high rainfall variability.

FIGURE 1 The amount of rain that falls in Australia varies from place to place, as this rainfall map shows.

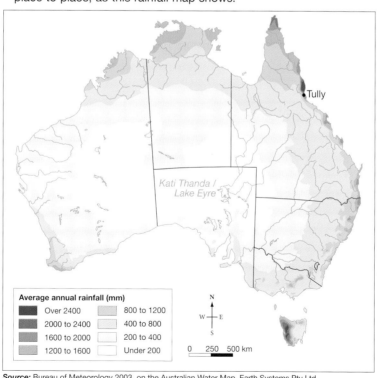

Source: Bureau of Meteorology 2003, on the Australian Water Map, Earth Systems Pty Ltd

3.4.3 Evaporation

Another problem for Australia is that most of its rainfall does not end up in rivers; much of it evaporates. Of all the water carried by the world's rivers, Australian rivers contain only one per cent of that total — even though Australia has five per cent of the world's land area. On average, only 10 per cent of our rainfall runs off into rivers and streams or is stored as groundwater. This figure drops to 3 per cent in dry areas and rises to 24 per cent in wetter places. The rest evaporates, is used by plants, or is stored in lakes, wetlands or underground storages. Areas in central Australia are very dry and, as a result, have high evaporation rates. Coastal areas have lower **evaporation** rates because they are close to the sea.

Relative humidity is a measure of the air's moisture content expressed as a percentage of the maximum moisture the air can contain at a certain temperature. Keep in mind that warm air can contain more moisture than cool air. Relative humidity does not measure the exact amount of moisture in the air because that depends on air temperature. For example, if Brisbane has a day of 30 °C and Melbourne has a day of 15 °C, and the relative humidity in both places is 60 per cent, there will be much more moisture in the air in Brisbane than in Melbourne.

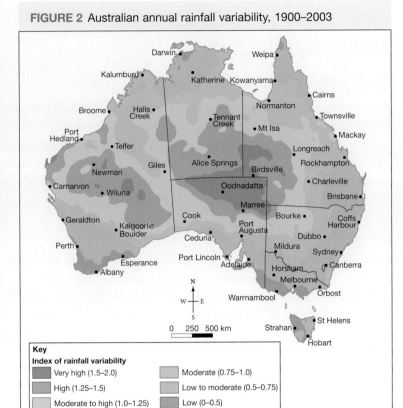

FIGURE 2 Australian annual rainfall variability, 1900–2003

Key

Index of rainfall variability

- Very high (1.5–2.0)
- High (1.25–1.5)
- Moderate to high (1.0–1.25)
- Moderate (0.75–1.0)
- Low to moderate (0.5–0.75)
- Low (0–0.5)

Source: MAPgraphics Pty Ltd, Brisbane

FIGURE 3 Average annual evaporation, Australia

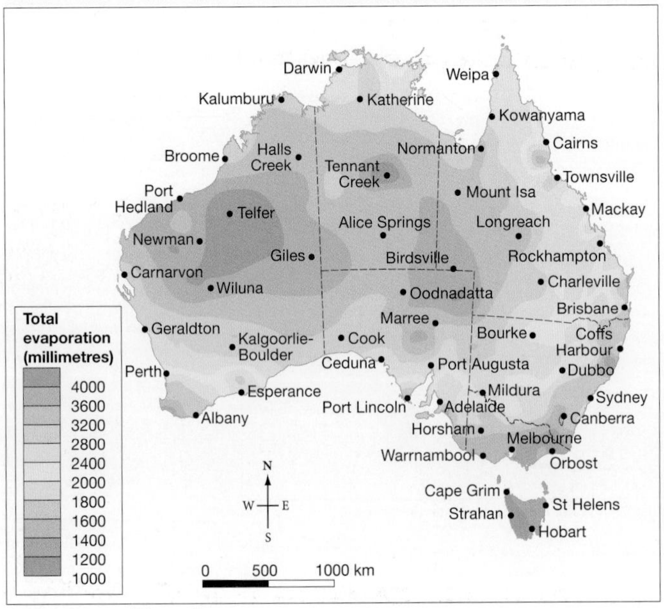

Total evaporation (millimetres)

4000
3600
3200
2800
2400
2000
1800
1600
1400
1200
1000

Source: Bureau of Meteorology

FIGURE 4 Average relative humidity across Australia

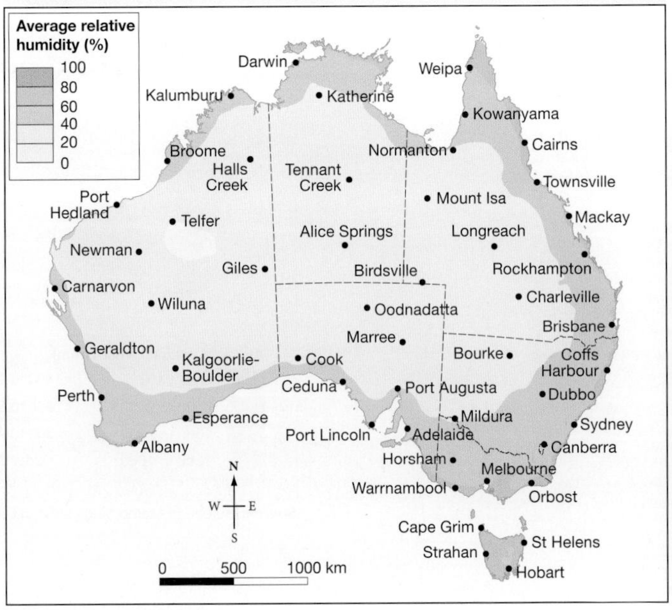

Average relative humidity (%)

100
80
60
40
20
0

Source: Bureau of Meteorology

Relative humidity tends to be higher in coastal regions, as is rainfall. Relative humidity tends to be higher in the parts of Australia that have very high rainfall, such as north Queensland and western Tasmania.

3.4 Activities

To answer questions online and to receive **immediate feedback** and **sample responses** for every question, go to your learnON title at www.jacplus.com.au. *Note:* Question numbers may vary slightly.

Remember

1. Which two regions receive the most rainfall in Australia?
2. Which region has the most variable rainfall?

Explain

3. What is rainfall variability?
4. Look at figure 2 and locate the *place* where you live. Does your area have high or low rainfall variability?
5. What is relative humidity? Use figure 4 to find the relative humidity of the *place* where you live.
6. Why does Australia have high evaporation rates?
7. Study the rainfall, humidity and evaporation maps. Fill in the missing word in the following statements in order to describe the *interconnections* between these features of our climate.
   Areas with low rainfall and low humidity tend to have a _____ evaporation rate.
   Areas with high rainfall and high humidity tend to have a _____ evaporation rate.

Discover

8. Find the *place* where you live on the map of Australia. Study the four maps in this section and complete a table like the one below. Compare where you live with another *place* in your state or territory and a *place* a long way from where you live.

	Average rainfall	Rainfall variability	Average evaporation	Relative humidity
Where I live: _____ _____				
Another place in my state/territory: _____ _____				
A place far from where I live: _____ _____				

9. Find out the average rainfall for Kati Thanda–Lake Eyre and the wettest locations and heaviest rainfalls in north-east Queensland and western Tasmania. Record the rainfall variability, evaporation and relative humidity. What differences are there between these locations?

Think

10. Australia has high evaporation rates and high rainfall variability. List all the ways that this *environment* makes water delivery to people a challenge.

 learn RESOURCES — ONLINE ONLY

Try out this interactivity: Hot and dry (int-3081)

 my**World**Atlas

Deepen your understanding of this topic with related case studies and questions.
❍ **Australia: weather and climate**

3.5 SkillBuilder: How to read a map

WHAT ARE MAPS AND WHY ARE THEY USEFUL?

Maps represent parts of the world as if you were looking down from above. Cartographers use colours and symbols on the map to show how features such as roads, rivers and towns are organised in a spatial way. Maps are useful to show features so that we have a deeper understanding of places.

Go online to access:

- a clear step-by-step explanation to help you master the skill
- a model of what you are aiming for
- a checklist of key aspects of the skill
- a series of questions to help you apply the skill and to check your understanding.

FIGURE 1 Essential map features

BOLTSS

B	Border — a box around the map to clearly show its extent
O	Orientation — a compass direction
L	Legend — a key to what the symbols and colours on the map stand for
T	Title — a clear indication of what the map is about or its theme
S	Scale — indicates distances on the map compared with the actual area being shown
S	Source — where possible, the information used to make the map should be sourced

learn on RESOURCES — ONLINE ONLY

Watch this eLesson: How to read a map (eles-1634)

Try out this interactivity: How to read a map (int-3130)

3.6 How is water used by people?

3.6.1 What is water used for?

There are three main uses made of water by all people: growing food, producing goods and electricity, and using it in the home. The amount of water consumed for each of these uses differs from one place to another. The problem remains that while the total amount of fresh water is fixed, the amount used per person is increasing.

It is interesting to look at water consumption on a global scale. With the global average at 1240 cubic metres per person, per year, some countries consume more water than others. Examples of countries that consume nearly twice as much as the global average are the United States and Thailand. Some countries that consume the least amount of water per person are Peru, Somalia and China.

Figure 1 shows that most of the world's water is used in agriculture, to grow food for the world's increasing population. This is especially the case in the drier parts of the world where there is not enough rainfall to grow crops or grass for animals. There is a strong interconnection between the amount of rainfall in a region and the amount of water used in agriculture.

It is interesting to see how this pattern varies in different countries. In some countries, the water used in agriculture and industry is greater than the amount of water used in homes for domestic use. In other places, people consume more water for domestic use than for either agriculture or industry.

myWorldAtlas   Deepen your understanding of this topic with related case studies and questions.
- **How is water used in Australia?**
- **North-west Europe**

FIGURE 1 Countries in the world differ in their use of water.

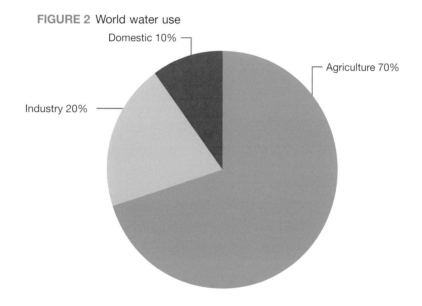

Main uses of water (in order of use)

Agriculture, industry, domestic/municipal

Industry, domestic/municipal, agriculture

Domestic/municipal, agriculture, industry

No data

Agriculture, domestic/municipal, industry

Industry, agriculture, domestic/municipal

Domestic/municipal, industry, agriculture

Difference between first and second uses is 10% or less

Source: Mekonnen, M.M. and Hoekstra, A.Y. 2011, 'National water footprint accounts: the green, blue and grey water footprint of production and consumption', Value of Water Research Report Series No. 50, UNESCO-IHE, Delft, the Netherlands

FIGURE 2 World water use

Domestic 10%
Industry 20%
Agriculture 70%

3.6.2 How is water used in Australia?

Agriculture is an important industry in Australia, and it is our thirstiest industry. It produces most of our food requirements and contributes enormously to Australia's export earnings.

Around 70 per cent of Australia's fresh water is used as irrigation for farming. Many crops are grown in dry areas where up to half the available water evaporates from the soil surface or seeps down too low into the ground for plant roots to reach it. Therefore, more water is applied than is actually needed by plants. In manufacturing industries, most water is used to produce food, beverages and paper.

In many areas in Australia where rainfall is limited or highly seasonal, farmers irrigate their crops with water stored in dams, with groundwater or with water from major rivers. Irrigation is a very important use of water in Australia. Most large-scale farming could not provide food for Australia's population without using water from rivers, lakes, reservoirs and wells.

There is high demand for irrigation water during summer when river flows are low, and low demand for irrigation water during winter when river flows are high. This reverses the natural pattern of river flow.

myWorldAtlas Deepen your understanding of this topic with related case studies and questions.
❯ **Mediterranean basin**

TABLE 1 Fresh water use in Australia

Types of use of fresh water	%
Agriculture • (pasture 35%) • (crops 27%) • (rural and domestic stock 8%)	70
Urban	12
Horticulture	10
Industry	3
Mining	2
Services	2
Hydro-electricity	1

TABLE 2 Fresh water used to irrigate different crops in Australia

Crop type	Water (gigalitres)	%
Livestock, pasture, grains and other agriculture	8795	56
Cotton	1841	12
Rice	1643	11
Sugar	1236	8
Fruit	704	5
Grapes	649	4
Vegetables	635	4

Note: One gigalitre = 1 000 000 000 litres or one thousand million litres or 400 Olympic-sized swimming pools

FIGURE 3 Australia is one of the most irrigated countries in the world.

3.6 Activities

To answer questions online and to receive **immediate feedback** and **sample responses** for every question, go to your learnON title at www.jacplus.com.au. *Note:* Question numbers may vary slightly.

Remember

1. What is most of the world's water used for?

Explain

2. Why might some countries use more water in industry than in agriculture or domestic use?
3. Use the data in tables 1 and 2 to draw two line graphs. Use the data from your graphs to describe how water is used in Australia. Which crops use the most water? Which use the least?

Discover

4. Research the Luangwa, Kafue and Zambezi rivers in Zambia, Africa. Use an atlas to trace the flow of the river and list some of the different water uses.
5. Study figure 1 and decide which of the following statements are true and which are false.
   (a) Australia uses most water for agriculture, then industry, then domestic/municipal.
   (b) Countries in North Africa use most water for industry, then domestic/municipal, then agriculture. (Use the **Regions** resource in the Resources tab).
   (c) Belarus uses most water for industry, then agriculture, then domestic/municipal.
   (d) Colombia uses most water for agriculture, then domestic/municipal, then industry.
   (e) Belize uses most water for industry, then agriculture, then domestic/municipal.
   (f) Malaysia uses most water for industry, then agriculture, then domestic/municipal.
Rewrite the statements that are false and make them true.

Think

6. Now use your atlas to look at patterns in figure 1 and compare them to a map that shows global wealth.
   (a) Name two countries with low wealth and high water use in industry.
   (b) Name two wealthy countries that do not have high water use in industry.
   (c) Can you write a general statement about wealth and water use? Add two exceptions to your general statement.
   (d) Discuss with students in a small group the fairness between wealth and water use. Should access to water be equal to all as a basic human right?

---

 learn on RESOURCES — ONLINE ONLY

Complete this digital doc: Regions (doc-17950)

---

my**World**Atlas — **Deepen your understanding of this topic with related case studies and questions.**
  ● **Three rivers in Africa**

3.7 SkillBuilder: Drawing a line graph

 online only

WHAT IS A LINE GRAPH?

A line graph displays information as a series of points on a graph that are joined to form a line. Line graphs are very useful to show change over time. They can show a single set of data, or they can show multiple sets which enables us to compare similarities and differences between two sets of data at a glance.

FIGURE 1 Water use in the Murray–Darling Basin

Water use in the Murray–Darling Basin, 1950–2000

Go online to access:

- a clear step-by-step explanation to help you master the skill
- a model of what you are aiming for
- a checklist of key aspects of the skill
- a series of questions to help you apply the skill and to check your understanding.

Source: http://www.waterforgood.sa.gov.au/rivers-reservoirs-aquifers/murray-darling-basin-plan/water-use-inthe-basin

3.8 How is water used by indigenous peoples?

Access this subtopic at **www.jacplus.com.au**

3.9 Does everyone have enough water?

3.9.1 The human right to water

The right to water is a human right that is protected by many international agreements, yet not everyone has access to this life-giving resource.

Everyone has the right to enough safe, accessible and affordable water for all their needs. Water is more important to survival than food. In hot conditions, a person can survive up to three weeks without food but only two or three days without water.

People need access to **improved drinking water**, yet over 660 million people use unclean drinking water. Water is also needed to cook food, to bathe, to wash dishes and clothes, and to flush toilets. However, with the global population increasing and a fixed amount of water on Earth, some regions are suffering **water scarcity**. Water scarcity occurs when the demand for water is greater than the available supply.

FIGURE 1 Global fresh water availability

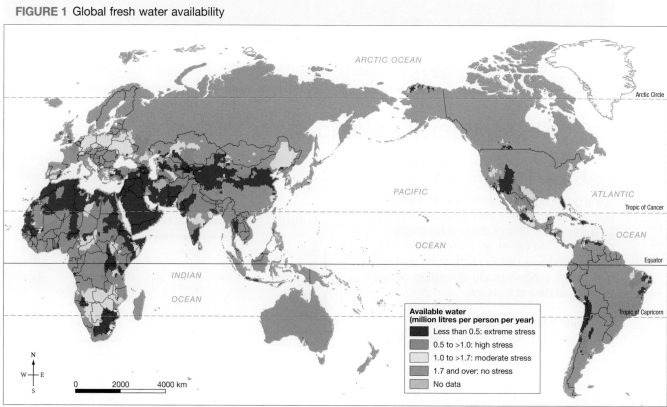

Available water (million litres per person per year)
- Less than 0.5: extreme stress
- 0.5 to >1.0: high stress
- 1.0 to >1.7: moderate stress
- 1.7 and over: no stress
- No data

Source: Spatial Vision

Ideally, each individual needs one cubic metre (1000 litres) of drinking water per year, about 100 cubic metres for other personal needs, and 1000 cubic metres to grow all the food that he or she consumes. **Water stress** occurs when there is not enough water available for all demands. A country with less than 1000 cubic metres of renewable fresh water per capita (per person) is under water stress.

3.9.2 Access to water

A major reason for so many people lacking access to safe water is the difference between where people live and where rain falls. Other reasons include water being used for agriculture and industry in regions where it is dry, and water being so polluted it cannot be used.

As climate change conditions take hold, it is estimated that by 2025, 1.8 billion people will be living in regions with absolute water scarcity, and two-thirds of the world's population could be living under water stressed conditions. The problem of lack of water is often worse in rural areas, so many people move from the countryside into towns and cities, hoping for a better water supply. These people are sometimes called water refugees. However, the water in some cities is also inadequate because it is in short supply or is very polluted.

3.9.3 The water carriers

People who do not have water at home have to travel to get water. Water is very heavy and difficult to carry. The burden of this water-fetching usually falls on women, who carry the heavy load on their head or back. For some people, the trip to a water supply and back can take hours each day. The average distance that women in Africa and Asia walk to collect water is six kilometres. The average weight they carry on their heads is about 20 kilograms — the usual weight of a suitcase taken on a flight. The World Health Organization estimates that over 40 billion work hours are lost each year in Africa alone, just collecting drinking water.

FIGURE 2 Some people have access to cleaner water than others.

Access to unimproved water: unprotected well or spring, water cart, tanker truck (13%)

Access to improved water: piped water to dwelling or gardens (54%)

Access to improved water: public tap or standpipe, tube well or borehole, protected dug well, protected spring, rainwater collection (33%)

FIGURE 3 Drought conditions can reduce water quantity and quality.

Use the **Burden of thirst** weblink in the Resources tab to watch a video about water scarcity in east Africa.

FIGURE 4 Women bear the burden of collecting water.

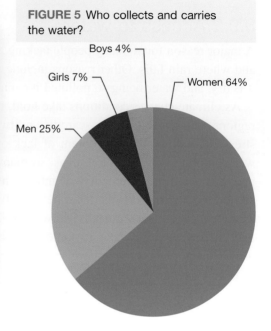

FIGURE 5 Who collects and carries the water?

Boys 4%
Girls 7%
Women 64%
Men 25%

3.9 Activities

To answer questions online and to receive **immediate feedback** and **sample responses** for every question, go to your learnON title at www.jacplus.com.au. *Note:* Question numbers may vary slightly.

Remember

1. Why is access to water a human right? Discuss this in small groups and report your ideas back to the class in a class discussion.
2. What is meant by water scarcity and water stress?

Explain

3. What is the difference between improved and unimproved water supplies?
4. Describe the impact on a country if it is under water stress or water scarcity.
5. There is a difference in water quality across Australia. Is your water supply 'improved'? Explain.
6. What might be meant by the term *water refugee*?

Discover

7. Study figure 1. Name three countries that are water stressed and experience water scarcity.
8. Geographers like to look at patterns over *space*. Find an atlas map showing population density and compare it with figure 1. Refer to countries in North Africa and West Asia in your answers. (Refer to the **Regions** resource in the Resources tab.)
   (a) Name three countries that have high population densities and are experiencing water scarcity.
   (b) Name countries with low population densities experiencing water stress or scarcity.
9. Use figure 1 to describe the water scarcity in North Africa.

Think

10. Study figure 5. Women are the main water carriers in *places* where there is water stress and scarcity. Discuss with another student how this would affect a woman's health, education, family life and food production. Draw a consequence map of your ideas.
11. What would be the impact on the lives of women and children if there was a well with clean water in every village or town? Discuss in small groups the fairness or otherwise of this situation. How could this be improved for women and children?

 **RESOURCES** – ONLINE ONLY

Complete this digital doc: Regions (doc-17950)

 myWorldAtlas **Deepen your understanding of this topic with related case studies and questions.**
◦ The Dead Sea — overcoming water scarcity
◦ Russia and Eurasia

3.10 How does access to water improve health?

3.10.1 How does dirty water affect health?

Everyone in the world has the right to an adequate supply of water. The right to water is linked to many other rights, including the right to food and to health.

More than 660 million people in the world have no access to clean water, and more than 2.3 billion people have no safe way of disposing of human waste. Lack of toilets means many people defecate in open spaces or near the same rivers from which they drink. It is estimated that 90 per cent of sewage in poor countries ends up flowing straight into rivers and creeks.

This is a unacceptable situation. Dirty water and lack of proper hygiene kill around 315 000 children around the world every year, most of them younger than five. People who are sick are often unable to work properly, to look after their families or to attend school, adding to the poverty cycle they may already be in. The diseases that can be passed on to people as a result of contaminated water include diarrhoeal diseases such as cholera, typhoid and dysentery. Malaria, a disease transmitted by mosquitoes, kills about a million people every year.

3.10.2 How can water-borne diseases be reduced?

People use different methods to treat the water they have collected. They can let it stand and settle, strain it through a cloth, filter it, add bleach or chlorine, or boil it. Some people do not treat their water at all.

When there is barely enough water to drink or to cook with, it is difficult for people to set aside water for washing hands and cleaning clothes. However, hygiene and sanitation are very important for health.

A number of aid groups, such as WaterAid, Water.org, CARE and A Glimmer of Hope, work on projects to improve sanitation and access to clean water. Washing hands, building cheap and effective toilets and teaching the community about good hygiene all help to reduce disease.

FIGURE 1 Deaths caused by poor sanitation and dirty water, 2012

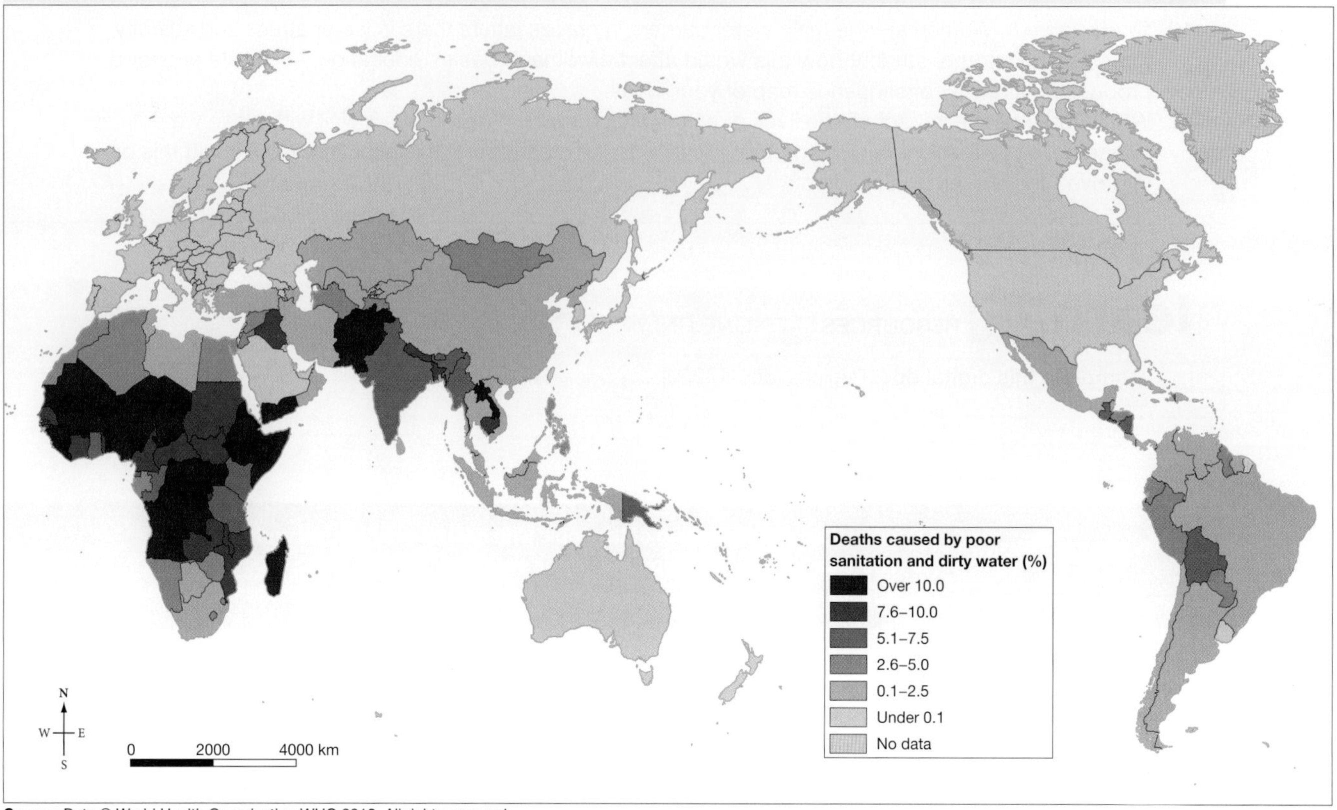
Deaths caused by poor sanitation and dirty water (%)
Over 10.0
7.6–10.0
5.1–7.5
2.6–5.0
0.1–2.5
Under 0.1
No data
N
W
E
S
0
2000
4000 km

Source: Data © World Health Organisation WHO 2012. All rights reserved

3.10.3 Sustainable Development Goals

Developed by the United Nations, the Sustainable Development Goals (SDGs) came into force on 1 January 2016. They are goals that aim to end all forms of poverty, fight inequalities and tackle climate change by the end of 2030. The SDGs build on the success of the Millennium Development Goals (MDGs) that were adopted from 2000–2015.

Goal 6 of the SDGs is to 'Ensure access to water and sanitation for all'. From 1990 to 2015 — including the 15 years of the Millennium Development Goals — the percentage of people who had access to clean water increased from 76 to 91 per cent.

Some of the targets (by 2030) for Goal 6 are to achieve:

- universal and equitable access to safe and affordable drinking water for all
- access to adequate and equitable sanitation and hygiene for all and end open defecation, paying special attention to the needs of women and girls and those in vulnerable situations
- improve water quality by reducing pollution, eliminating dumping and minimising release of hazardous chemicals and materials, halving the proportion of untreated wastewater and substantially increasing recycling and safe reuse globally
- substantially increase water-use efficiency across all sectors and ensure sustainable withdrawals

FIGURE 2 Collecting water that is unsafe to drink

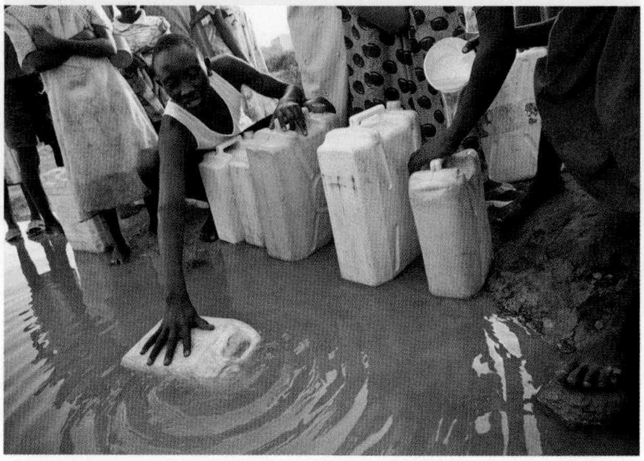

and supply of freshwater to address water scarcity and substantially reduce the number of people suffering from water scarcity
- protect and restore water-related ecosystems, including mountains, forests, wetlands, rivers, aquifers and lakes (this goal is by 2020).

The maps shown in figures 3 and 4 show the achievements of the MDGs in relation to water and sanitation goals. The SDGs aim to build on these achievements.

3.10.4 Case study: Nigeria

CASE STUDY

Community-led total sanitation (CLTS), Nigeria

WaterAid went to a village called Olorioko in the state of Ekiti to see if they could improve the sanitation. When they arrived, there were very high rates of illness and death due to water-related diseases. The people in the village used the bush near their houses as their toilet.

CLTS leaders developed a relationship with the villagers and taught them how faeces can enter their food and make them sick. Once this was understood, the people wanted to change their practices so this would no longer happen. Action plans were drawn up and eventually the villagers created clean water points, built simple but effective toilets, and were given lessons in sanitation.

The health of the entire village has improved and they are also increasing their wealth. The CLTS project, which started in Bangladesh some years ago, has been a success and is spreading throughout Nigeria.

Use the **Health is wealth** weblink in the Resources tab to learn about a community-led sanitation project.

FIGURE 3 MDG progress on access to water

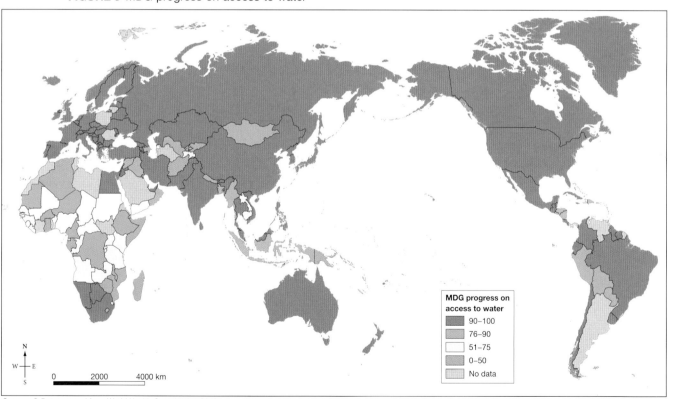

MDG progress on access to water
- 90–100
- 76–90
- 51–75
- 0–50
- No data

Source: © Data sourced from World Health Organisation 2012

FIGURE 4 MDG progress on access to sanitation

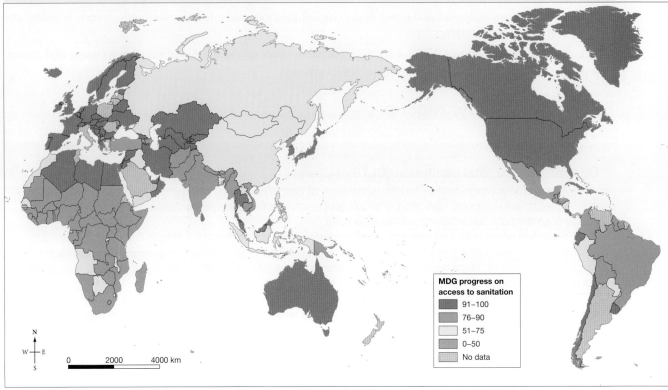

MDG progress on access to sanitation

- 91–100
- 76–90
- 51–75
- 0–50
- No data

Source: World Health Organisation / UNICEF Joint Monitoring Program JMP for Water Supply and Sanitation

FIGURE 5 Location of Nigeria in Africa

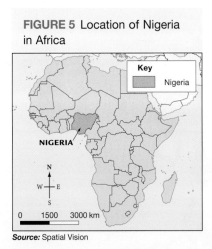

Key
Nigeria

NIGERIA

Source: Spatial Vision

3.10 Activities

To answer questions online and to receive **immediate feedback** and **sample responses** for every question, go to your learnON title at www.jacplus.com.au. *Note:* Question numbers may vary slightly.

Remember

1. How many people in the world do not have access to clean water or sanitation? How do the SDGs aim to improve this situation?
2. Describe Nigeria's location in Africa and in relation to other countries and *places*.
3. What does CLTS stand for?

Explain

4. Study figure 1. Look at sub-Saharan and North Africa. (Use the **Regions** weblink in the Resources tab.)
   (a) Name three coastal and two landlocked countries in these regions.
   (b) How many countries in this region suffer deaths caused by poor sanitation and dirty water?
   (c) Name the continents and regions that have the fewest deaths.
   (d) Why do you think this *spatial* pattern exists? (*Hint:* Look at maps in your atlas that show wealth.)
5. Explain what the community-led total sanitation project aims to do.
6. Use the information in the maps to explain the success of the MDGs in achieving access to water and sanitation. Explain this for each world region. Use the **WHO/UNICEF** weblink in the Resources tab to see an interactive world map showing specific figures for water and sanitation.

Predict

7. What might happen to people's health in North and sub-Saharan Africa if access to water and sanitation is not improved?
8. A number of aid agencies are working in countries and regions to improve access to sanitation and clean water. Choose one of those listed in this topic and find out more about what they are doing. How will their work make a difference to the living conditions of the people they are helping?

Think

9. Work in groups of three or four. Use the data and facts in these pages to plan a day of promoting knowledge about this issue at your school. Use the links available at the **UN SDG** weblink in the Resources tab which includes information on programs. Make particular reference to North and sub-Saharan Africa, and find out what is being done by aid organisations to improve the situation in these regions. Plan a video presentation that is interesting and catchy and will help people understand the action needed to improve access to clean water and sanitation. Use video and video editing programs and internet research in your planning.

 learn on RESOURCES — ONLINE ONLY

📄 **Complete this digital doc:** Regions (doc-17950)

🔗 **Explore more with these weblinks:** Health is wealth, UN SDG, WHO/UNICEF

my**World**Atlas   **Deepen your understanding of this topic with related case studies and questions.**
⊙ **World: health**

3.11 What is virtual water?

3.11.1 Virtual water

The water we consume is not just what we use in cooking, drinking, washing, flushing or in the garden. Water is used to manufacture everything we use: mobile phones, toys, cars and newspapers. This **virtual water** needs to be considered in our **water footprint** — that is, the water used to produce all our goods and services.

Virtual water is also known as embedded water, embodied water or hidden water. It includes all the water used to produce goods and services. Food production uses more water than any other production.

Hidden in a cup of coffee are 140 litres of water used to grow, produce, package and ship the beans. That is roughly the same amount of water used by an average person daily in Australia for drinking and household needs. There is a lot of water hidden in a hamburger too: 2400 litres. This includes the water needed to grow the feed for the cattle over a number of years, to grow wheat for the bread roll, to grow all the other ingredients in the hamburger, and to process all the food.

Virtual water varies from food to food. For example, it takes about 3400 litres of water to grow one kilogram of rice, whereas it takes 200 litres to grow one kilogram of cabbage. Regions that are water stressed and that export food and other products (such as Australia and some countries in Africa and Asia) are also effectively exporting their precious water in these goods.

A country that imports rice, rather than growing it locally, therefore saves 3400 litres of water for every kilogram it imports. Some countries (such as Japan) have very little land on which to grow food; other countries have very few cubic metres of renewable water per person. Singapore, for example, has only about 130 cubic metres per person. Both types of countries survive by virtual water imports: they import food rather than attempt to grow and produce all their food themselves. This means that small, wealthy countries can import food that needs a lot of water to produce, and export products that need little water to produce. This makes water available for other domestic purposes such as drinking and cooking.

TABLE 1 The water used to grow food varies from product to product

Food item	Unit	Global average water (litres)
Apple or pear	1 kg	700
Barley	1 kg	1300
Banana	1 kg	860
Beef	1 kg	15 500
Beer (from barley)	250 mL	75
Bread (from wheat)	1 kg	1300
Cabbage	1 kg	200
Cheese	1 kg	5000
Chicken	1 kg	3900
Chocolate	1 kg	24 000
Coconut	1 kg	2500
Coffee (roasted)	1 kg	21 000
Cotton shirt	1	2700
Cucumber or pumpkin	1 kg	240
Dates	1 kg	3000
Eggs	1	200
Goat meat	1 kg	4000
Groundnuts (in shell)	1 kg	3100
Hamburger	1	2400
Lamb	1 kg	6100
Leather	1 kg	16 600
Lettuce	1 kg	130
Maize	1 kg	900
Mango	1 kg	1600
Millet	1 kg	5000
Milk	250 mL	250
Olives	1 kg	4400
Orange	1	50
Paper	1 A4 sheet	10
Peach or nectarine	1 kg	1200
Pork	1 kg	4800
Potato	1 kg	4800
Rice	1 kg	3400
Soybeans	1 kg	1800

Food item	Unit	Global average water (litres)
Sugar (from sugar cane)	1 kg	1500
Tea	250 mL	30
Tomato	1 kg	180
Wheat	1 kg	1300
Wine	125 mL	120

The major exporters of virtual water are found in North and South America (the United States, Canada, Brazil and Argentina), south Asia (India, Pakistan, Indonesia, Thailand) and Australia. The major virtual water importers are North Africa and the Middle East, Mexico, Europe, Japan and South Korea.

3.11.2 What is a water footprint?

The water footprint of an individual or country is the total volume of fresh water that is used to produce the goods and services consumed by the individual or country. It includes the use of:
- blue water (rivers, lakes, **aquifers**)
- green water (rainfall used for crop growth)
- grey water (water polluted after agricultural, industrial and household use).

FIGURE 1 Average water footprints

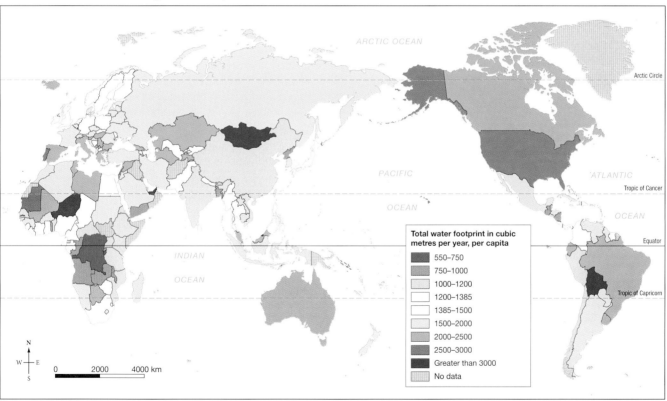

Source: waterfootprint.org

Not all goods consumed in one particular country are produced in that country — some foods and products are imported. Therefore, the water footprint consists of two parts: use of domestic water resources and use of water outside the borders of the country.

In the United States, the average water footprint per year per capita is 2842 cubic metres, which is enough to fill an Olympic swimming pool. In China, the average water footprint is 1071 cubic metres per year. The figure for Australia is 2315 cubic metres per year.

Japan, with a footprint of 1379 cubic metres per person per year, has about 65 per cent of its total water footprint outside the borders of the country, meaning a lot of its water is imported in the form of consumer goods and food. On the other hand, only about seven per cent of the Chinese water footprint falls outside China.

3.11 Activities

To answer questions online and to receive **immediate feedback** and **sample responses** for every question, go to your learnON title at www.jacplus.com.au. *Note:* Question numbers may vary slightly.

Remember

1. Explain the difference between virtual water and a water footprint.
2. Outline the differences between blue, green and grey water.

Explain

3. Refer to figure 1. Describe the patterns you notice over *space* of countries with (i) very high and high water footprints and (ii) very low and low water footprints.
4. Use a blank world map and choose two colours to show the main importers and exporters of virtual water. Describe the *spatial* patterns of the map you have drawn.

Discover

5. Study table 1. Choose three meat, five grain, two dairy, two non-food, four fruit, four vegetable and two processed products from the list. Create a bar graph to show how much water is used to produce a vegetarian diet and a meat-based diet. Which diet uses more water?
6. Use the **Just add water** weblink in the Resources tab to listen to an audio program about the water footprint in food production.
   (a) What is the relationship between water-stressed countries and food production?
   (b) Give an example where the water footprint figure is in conflict with the opinion of farmers.
   (c) Detail a product produced in Wodonga that includes both virtual and domestic water.

Think

7. Conduct a debate on the following statement: 'People should eat less meat in order to consume less water.'
8. Compare the data for North Africa and West Asia (the Middle East) in figure 1 and the maps in subtopics 3.2, 3.6 and 3.9. (Also use the **Regions** resource in the Resources tab.) Write three summary statements that describe the amount of rainfall, water use (including water footprints) and water availability for these two regions. How do these patterns compare with Australia?
9. Describe how water footprints *interconnect places* often far away from each other.

learn **on** RESOURCES — ONLINE ONLY

- **Try out this interactivity:** Unreal (int-3080)
- **Explore more with this weblink:** Just add water
- **Complete this digital doc:** Regions (doc-17950)

 my**World**Atlas **Deepen your understanding of this topic with related case studies and questions.**
❯ **Our water footprint**

3.12 How does water quality change?

3.12.1 Polluted rivers

Water quality can affect health in many ways. Rivers and streams act as drainage systems and, when it rains, water transports rubbish, chemicals and other waste into drains and, eventually, rivers.

Different pollutants — faeces (human and animal), food wastes, pesticides, chemicals and heavy metals— can come from industrial wastewater, domestic sewage, cars, gardens, farmland, mining sites and roads, and flow into waterways.

Some countries, cities and local areas are better than others at providing services and enforcing laws to prevent pollutants from entering water. Some of the worst polluted rivers and lakes in the world include rivers and aquifers in China (such as the Songhua River and the Yellow River), the Citarum River in Indonesia, the Yamuna and Ganges rivers in India, the Buriganga River in Bangladesh and the Marilao River in the Philippines.

3.12.2 How clean is your river?
Fieldwork investigating waterways

Some schools are located close to a waterway, even if it is a highly modified one like a concrete drain. Conducting fieldwork at a local waterway will help you to better understand national and global issues.

The aim is to investigate the physical properties of a river or creek at various points along its length, and to make observations of the water quality and any evidence of human impact. Does the quality of the water change between upstream and downstream sites? Are there human factors that can account for these changes? Differences may be more obvious if the waterway passes through a built-up area or a farm.

The following activities should be undertaken at each site and recorded on paper or directly onto a database on a laptop computer or other mobile device. Use a map and camera to record observations about the surroundings of each site, especially the amount of vegetation and possible human impact. Use GPS to record and map your location at each site.

Measuring river width

Stretch a tape measure 20 centimetres above the water from one bank to the other, measuring from where the dry bank meets the water. Take your reading directly above the tape at several locations, and calculate an average.

Measuring the water depth

While the tape measure is stretched across the river or creek, use a metre rule to measure the depth. Record the depth every 50 centimetres (or 30 centimetres if the creek is small). Make sure the ruler only just touches the riverbed, and record each measurement as it is taken.

Temperature

Aquatic plants and animals have a particular temperature range in which they can survive. High water temperatures can result in reduced oxygen available for plants and animals. It is useful to compare temperature readings with **biodiversity** counts to investigate this relationship. Place the bulb of a thermometer in the water for five minutes and record the result.

pH

We use pH to measure the acidity or alkalinity of water on a scale of 1 to 10. Drinking water should have a pH reading of around 6. A reading either side of this may indicate that water is polluted. You can test pH by taking a sample of water and using pH paper or chemical reagents.

Turbidity

Water that lets little sunlight through is said to be turbid. Turbidity is the amount of suspended sediment in water — sediments such as clay, silt, industrial waste or sewage. A Secchi disc is used to measure turbidity. This can be made using an ice-cream lid, string, a weight and black paint.

Lower the disk into the water, making sure there are no waves or ripples, cloud or glare. Do not wear sunglasses when making the reading. When it is only just visible, record the depth in centimetres. The lower the number, the greater the turbidity.

Salinity

Salinity measures the amount of salt in the water. To measure salinity, you will need to use an electrical conductivity meter, or EC meter, which can be bought from science equipment suppliers.

Biodiversity

Biodiversity in water is studied by taking a small area of water and investigating the number and diversity of animal species. It is measured by ponding or water sampling. Choose a number of sites in the water and ensure that they contrast with each other: there should be clear, muddy, deep, shallow, moving and still sites, for example.

The materials needed for ponding include a fine net; a white plastic ice-cream container; a magnifying glass; a notebook and pencil for recording the number and variety of species; a camera for photographing specimens; and an identification chart or book.

Aesthetics — what the water looks like

Another measurement of water quality is to grade the appearance of the water. If an area of water is appealing to look at, it is considered aesthetically appealing. Observe aspects such as colour, odour, and the presence of algae, surface film or oil slicks. Ratings from 1 (excellent) to 5 (extremely poor) can be used and recorded. Photographs and field sketches of the sites are also useful.

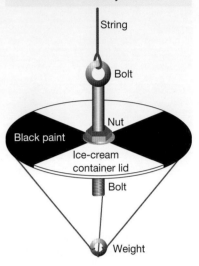

FIGURE 2 You can create a simple Secchi disc and use it to measure turbidity.

String
Bolt
Nut
Black paint
Ice-cream container lid
Bolt
Weight

FIGURE 3 Students measuring river water quality

3.12 Activities

Remember

1. How do rivers get polluted?
2. What is an aquifer?

Explain

3. You will need to organise and present the information you collected in a field report. This can be presented in Google Maps, or as a Prezi presentation, web page or multimedia report.
    Determine some headings for your report such as a main title; introduction and background; data and findings; and conclusion. Arrange your data and choose the best way to represent it, whether in the form of graphs, tables, two maps, overlays, or photo displays and diagrams. Write an analysis of your data by describing it in detail. Draw conclusions and provide an evaluation of the field trip.
4. You should alert your local council to your findings by inviting someone to a presentation or sending a link to your findings to an appropriate representative. Is there action that needs to be taken to improve this waterway?

Discover

5. Work in groups to investigate one of the other polluted rivers mentioned in the text. Show where it is located on a map. What does it look like? What has caused the pollution? Is anything being done to improve conditions?

learnon RESOURCES — ONLINE ONLY

Complete this digital doc: Regions (doc-17950)

3.13 How can water be managed?

3.13.1 Managing our water supply

More water cannot be created, but it can be managed better. With a growing global population, and the predicted changes due to climate change, the pressure on this finite resource requires a number of solutions.

Introducing effective water management can be a challenge at any scale, whether local, national or global. It needs the cooperation of all users, including farmers, industry, individuals, and upstream and downstream people in different countries or different states. With all the competing demands on water, management is often easier to approach at a local scale.

Because agriculture uses the greatest amount of water, it makes sense to make irrigation systems more efficient. The aim is to get more production for every drop of water used. Some irrigation systems waste up to 70 per cent of their water through leaks and evaporation, so changing the irrigation method can save water. Other management practices include recycling, using desalinated water and using stormwater.

3.13.2 Managing water across borders

About 260 drainage basins across the world are shared by two or more countries. Thirteen river basins are shared by five or more countries. Depending on their location in the catchment, some countries can suffer reduced access to water because of other countries' usage. This shows the *interconnection* between places — what happens in one place affects another. Diverting rivers, building dams, taking large amounts of water out for irrigation, and creating pollution can all lead to conflict between countries, states and political groups.

Country disputes have occurred in the Nile Basin in North Africa, along the Mekong River in Asia, the Jordan River Basin in West Asia (the Middle East) and along the Silala River in South America. This can also happen within a country, which has happened with the Murray–Darling Basin in Australia, across four states and one territory.

FIGURE 1 Kurnell desalination plant in southern Sydney provides water to the city's population. It is powered by wind energy produced in Canberra.

FIGURE 2 The Jordan River passes through three countries, and has tributaries in the north that flow in from another two countries.

Key
— Border
········ Disputed border

MEDITERRANEAN
SEA

LEBANON

Hasbani River

River

Banias

Hulah Valley

SYRIA

Sea of Galilee (Lake Tiberias)

Yarmouk River

River

Jabbok

River

West Bank

JORDAN

ISRAEL

Dead Sea

N
W — E
S

0   40   80 km

Source: Spatial Vision

Some countries sign international agreements or treaties to try and share water between nations. These include the Rhine and Danube rivers in Europe, the Nile River in North Africa, the Ganges and Brahmputra rivers in Asia and the Parana River in South America.

3.13.3 Managing water within a country: the Murray–Darling Basin

The Murray–Darling Basin (MDB) is Australia's largest catchment area, covering 14 per cent of the country's total landmass. It stretches across four Australian states (Queensland, New South Wales, Victoria and South Australia) and the Australian Capital Territory, and includes 23 separate river catchments. This is a changed system, in which water is taken out all along its length for storage in dams and use in irrigation. During long periods of drought, little water reaches the Murray mouth, often causing it to close. The Queensland floods in the summers of 2011 and 2012 provided the largest river flows for the MDB in many years.

Over the years the management of this river system was the responsibility of each of the four states and the territory, resulting in a lot of conflict. In 2008, the Australian Government took control of the MDB and prepared plans for how it should be managed. There is still controversy over the decisions being made.

FIGURE 3 Map of the Murray–Darling Basin. The Australian Government has been managing the basin.

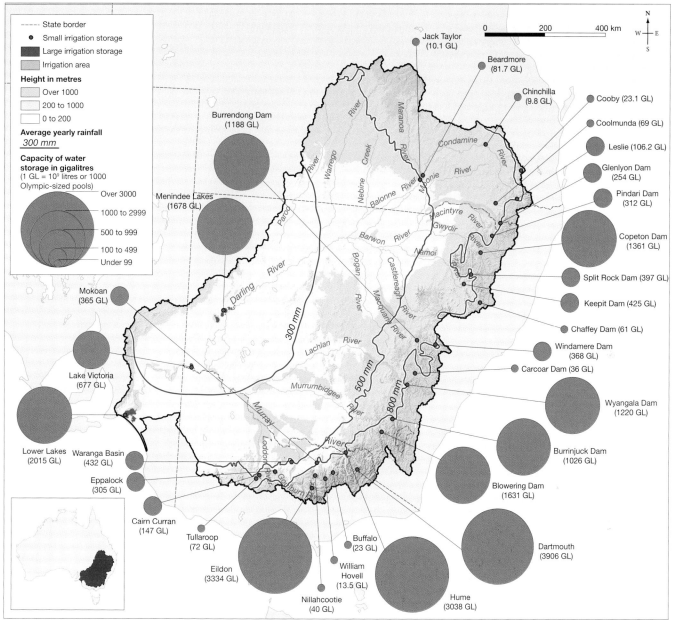

Source: Spatial Vision

Other examples of water basins that are shared by states within a country are the Colorado River and Ogallala Aquifer in the United States, and the Kaveri River in India. The Great Artesian Basin in Australia shares groundwater between Queensland, New South Wales, South Australia and the Northern Territory.

3.13.4 Other water management solutions for Australia

Over recent years, and especially after prolonged droughts, many different solutions have been suggested to solve our water problems. Some of these seem impractical, such as towing icebergs from Antarctica; others have generated much discussion, such as fixing all the leaking pipes in towns, cities and outback bores. Each of the suggestions has to be considered in light of various factors: cost, impact on people, impact on the environment, technology and politics.

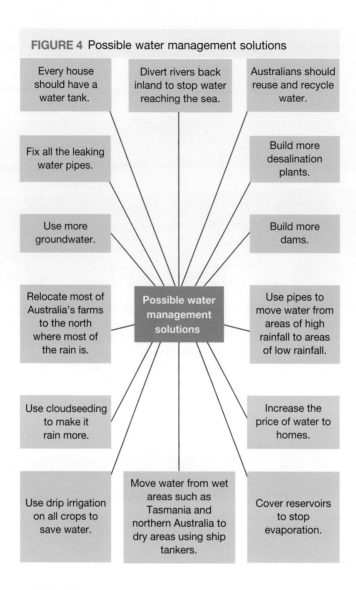

FIGURE 4 Possible water management solutions

Every house should have a water tank.

Divert rivers back inland to stop water reaching the sea.

Australians should reuse and recycle water.

Fix all the leaking water pipes.

Build more desalination plants.

Use more groundwater.

Build more dams.

Relocate most of Australia's farms to the north where most of the rain is.

Possible water management solutions

Use pipes to move water from areas of high rainfall to areas of low rainfall.

Use cloudseeding to make it rain more.

Increase the price of water to homes.

Use drip irrigation on all crops to save water.

Move water from wet areas such as Tasmania and northern Australia to dry areas using ship tankers.

Cover reservoirs to stop evaporation.

3.13.5 Managing water use at home

About 50 per cent of household water in Australia is used in the garden. Inside the house, approximately 80 per cent of water used is in the shower, toilet and laundry. It is predicted that our growing Australian population will, by 2051, need nearly twice as much water as we do now.

Policy changes such as water restrictions can reduce water usage. Some restrictions include bans on hosing down driveways, washing cars with hoses and watering private lawns, and cutting back sprinkler use during the day.

Tips to save water
In the house

- Take shorter showers.
- Turn off taps firmly and fix any leaks.
- Use water-efficient shower heads.
- Install a dual-flush toilet or adaptor.
- Use water-efficient appliances and use them only when they are full.
- Keep a jug of cold water in the fridge so you don't need to run the tap until the water is cold enough to drink.
- When replacing appliances, choose water-efficient models (AAA or AAAA rating).

In the garden
- Plant local native plants, which need only rainfall.
- Group plants with similar watering needs together and water them together.
- Use mulch on the garden beds to stop soil drying out — evaporation can be reduced by up to 70 per cent this way.
- Use a trigger nozzle on your hose.
- Install a timer on your outdoor taps.
- Use trickle irrigation systems rather than sprinklers for garden beds. They direct water where it is needed, and less water is lost to evaporation and wind-drift.
- Use a pool cover when the pool is not being used.

The long drought experienced by Australia between about 2003 and 2010 forced many governments to offer rebates on purchases of water-saving products. Items and services included buying and installing water tanks and grey water systems; dual flush toilets; water-saving shower heads and water-efficient washing machines. People were encouraged to spend money on these in order to save water.

3.13 Activities

To answer questions online and to receive **immediate feedback** and **sample responses** for every question, go to your learnON title at www.jacplus.com.au. *Note:* Question numbers may vary slightly.

Remember

1. Why is it difficult to manage water when the water supply crosses country or state borders?

Discover

2. Conduct some further research to find out how water is being managed along the Jordan River in West Asia (the Middle East). (Use the **Regions** resource in the Resources tab.)
3. Investigate the use of desalination in countries in West Asia.
4. Find out about the Murray–Darling Basin plan and how it will help manage water.

Predict

5. If water were oil, leaking pipes would be fixed immediately. However, there is still a perception that water is not as valuable as oil, so the same investment is not made in it. Imagine you work for an advertising company and have to convince your audience that water is more valuable than oil. Film your advertisement and use a video editing tool to create music and voiceovers. Present this to your class and your school.

Think

6. Use the **Catchment detox** weblink in the Resources tab to play the catchment-in-crisis game. See if you can manage the catchment so that agricultural, personal, industrial and environmental needs are all met. What might be the best water solutions for Australia?
7. Study the list of possible water solutions shown in figure 4 and choose the ones that are most likely to work. You will work in pairs to conduct some research about one of the proposed methods. There will be advantages and disadvantages for each. You will need to use the internet and libraries to find your information, and you will need to find out:
   - how the solution will be carried out, including the technology that might be used
   - which *places* in Australia are most likely to be involved and why (this might include rainfall data, Google images or maps, and photographs)
   - how much it will cost and who will pay
   - what impact the solution will have on the environment
   - what impact there will be on people
   - whether there are any political implications
   - where the solution will work best (shown on a map).

Presenting your information

Any data collected needs to be presented in an appropriate way. Climate data could be represented as a climate map; a satellite image or photograph could be annotated with notes; models or plans of the solution could be drawn.

Sharing your information

The class will need to share all the information found so that a decision can be made about the most probable solutions. Share via a presentation (such as a Prezi), a class wiki or blog.

Making a decision

Conduct a class vote to remove five of the 14 solutions immediately after the information sharing. Do this by writing each solution on a board and voting on each one. You will be left with nine water management solutions.

Diamond ranking activity

Use the remaining nine solutions to complete a diamond ranking. Make a copy of the nine solutions and write these on separate cards or sticky labels. Individually, use the information you shared about the solutions to rank the solutions from the most viable to the least viable from the top of the chart. It is often easier to work on the two extremes, top and bottom, and then continue working from there.

   Once you have your ranking, work in groups of four. Explain why you chose the ranking and see whether each of you can agree on the same ranking. Work together as a class and discuss the solutions again and see if you can arrive at a class ranking.

Conclusion

Write a short report, and include in it the final ranking of water management solutions for Australia. Include the discussions and explanations in your report.

   Now write a letter to your local state or territory water authority, outlining what you consider to be the three best management solutions for Australia and why.

8. Conduct a debate on the following statement: 'There is enough water for all purposes if it is managed well.'

 RESOURCES — ONLINE ONLY

Try out this interactivity: Ways forward (int-3082)

Complete this digital doc: Regions (doc-17950)

Explore more with this weblink: Catchment detox

---

 myWorldAtlas

Deepen your understanding of this topic with related case studies and questions.
- **Salisbury Council — Aquifer storage, transfer and recovery**
- **Murray–Darling Basin**

3.14 Review

 online only

3.14.1 Review

The Review section contains a range of different questions and activities to help you revise and recall what you have learned, especially prior to a topic test.

3.14.2 Reflect

The Reflect section provides you with an opportunity to apply and extend your learning.
Access this subtopic at **www.jacplus.com.au**

TOPIC 4
Too much, too little

4.1 Overview

Numerous **videos** and **interactivities** are embedded just where you need them, at the point of learning, in your learnON title at www.jacplus.com.au. They will help you to learn the content and concepts covered in this topic.

4.1.1 Introduction

Every person on the planet interacts with the weather on a daily basis. Sometimes we feel hot, sometimes we feel cold. The constant change in the weather also affects the environment.

Weather influences the level of precipitation experienced in different places. If there is too little rain, drought can develop, sometimes producing heatwaves — days of dry, hot weather. If there is too much rain, flooding will occur. These extreme weather events have many effects, which can be more severe in certain parts of the world. While we cannot control these events, we can learn from the past and plan to minimise their impact in the future.

Starter questions

1. Describe how the weather affects your everyday life.
2. List three things you would expect to happen if an area did not get its usual rainfall for a long time.
3. List three things you would expect to happen if an area suddenly got much more than its usual rainfall.
4. Think of an extreme weather event you have experienced or heard about in your life. Explain what happened and the effects this event had on you or others and the surrounding environment.

Flooding along Cooper Creek, Queensland

4.2 What is weather?

4.2.1 How does weather change?

Our Earth is surrounded by a band of gases called the atmosphere. It protects our planet from the extremes of the sun's heat and the chill of space, making conditions just right for supporting life. The atmosphere has five different layers. The layer that starts at ground level and ends about 16 kilometres above Earth is called the troposphere. Our weather is the result of constant changes to the air in the troposphere. These changes sometimes cause extreme weather events.

Droughts, floods, cyclones, tornadoes, heatwaves and snowfalls — even cloudless days with gentle breezes — all begin with changes to the air in the troposphere. The five main layers in the Earth's atmosphere all differ from one another. For example, the troposphere contains most of the **water vapour** in the atmosphere. As a result, this layer has an important link to precipitation.

All weather conditions result from different combinations of three factors:
• air temperature
• air movement
• the amount of water in the air.
The sun influences all three.

FIGURE 1 Australia experiences a diversity of weather, which has a major effect on how we live.

(a)

(b)

(c)

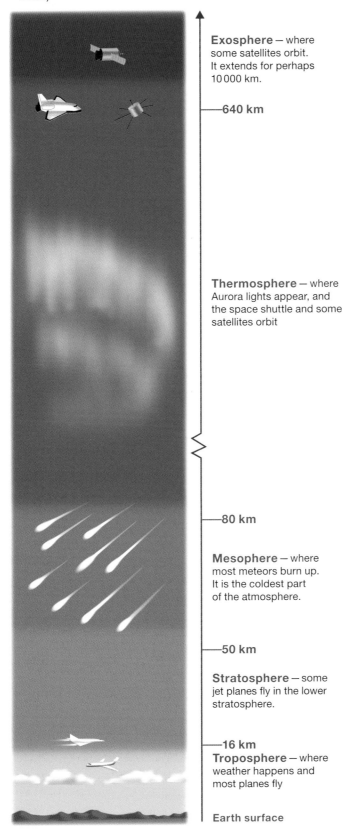

FIGURE 2 Structure of the Earth's atmosphere (not to scale)

Exosphere — where some satellites orbit. It extends for perhaps 10 000 km.

— 640 km

Thermosphere — where Aurora lights appear, and the space shuttle and some satellites orbit

— 80 km

Mesosphere — where most meteors burn up. It is the coldest part of the atmosphere.

— 50 km

Stratosphere — some jet planes fly in the lower stratosphere.

— 16 km
Troposphere — where weather happens and most planes fly

Earth surface

First, the sun heats the air. It also heats the Earth's surface, which in turn heats the air even more. How hot the Earth's surface becomes depends on the season and the amount of cloud cover.

Second, the sun causes air to move. This is because the sun heats land surfaces more than it heats the oceans. As the warm air over land gets even warmer, it expands and rises. When hot air rises, colder air moves in to take its place.

Third, the sun creates moisture in the air. The heat of the sun causes water on the Earth's surface to **evaporate**, forming water vapour. As this water vapour cools, it condenses, forming clouds. It may return to Earth as rain, dew, fog, snow or hail.

At times these three factors — temperature, air movement and water vapour — can create extreme weather events. Very high air temperatures influence heatwaves; rapidly rising air plays a part in the formation of cyclones; and excess rain can create flooding.

4.2.2 Weather and climate: what is the difference?

Weather is the day-to-day, short-term change in the atmosphere at a particular location. Extreme weather events are often described as unexpected, rare or not fitting the usual pattern experienced at a location.

Climate is the average of weather conditions that are measured over a long time. Places that share the same type of weather are said to lie in the same climatic zone. Because of the size of the Australian continent, its climate varies considerably from one region to another.

FIGURE 3 Climatic zones of Australia

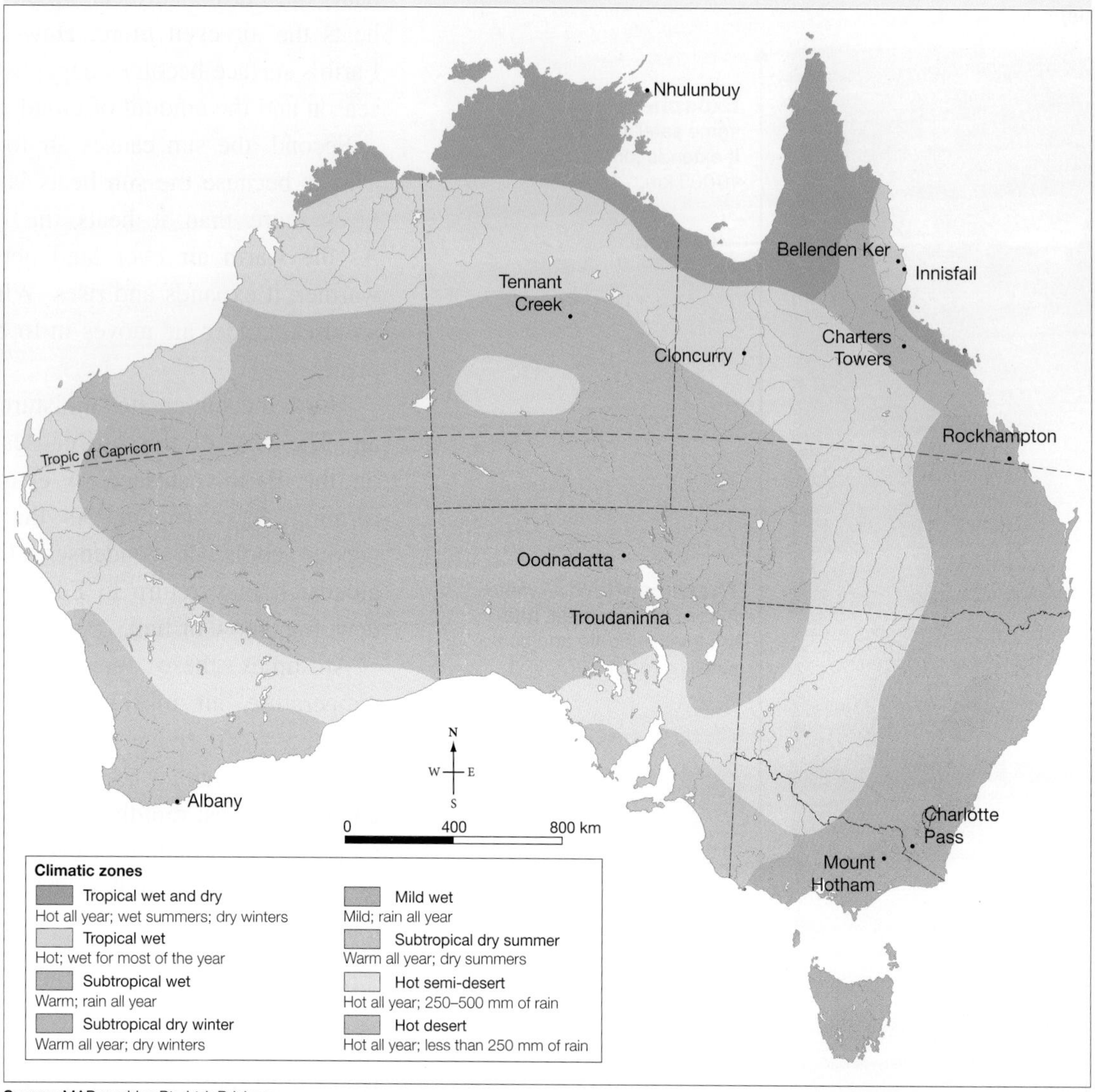

Nhulunbuy
Bellenden Ker
Innisfail
Tennant Creek
Cloncurry
Charters Towers
Rockhampton
Tropic of Capricorn
Oodnadatta
Troudaninna
Albany
Charlotte Pass
Mount Hotham
N
W
E
S
0
400
800 km
Climatic zones
Tropical wet and dry
Hot all year; wet summers; dry winters
Tropical wet
Hot; wet for most of the year
Subtropical wet
Warm; rain all year
Subtropical dry winter
Warm all year; dry winters
Mild wet
Mild; rain all year
Subtropical dry summer
Warm all year; dry summers
Hot semi-desert
Hot all year; 250–500 mm of rain
Hot desert
Hot all year; less than 250 mm of rain

Source: MAPgraphics Pty Ltd, Brisbane

4.2 Activities

To answer questions online and to receive **immediate feedback** and **sample responses** for every question, go to your learnON title at www.jacplus.com.au. *Note:* Question numbers may vary slightly.

Remember

1. What is the name of the layer of the atmosphere where all Earth's weather happens?
2. Define the term troposphere.
3. In which levels of the atmosphere are the following features found?
   (a) Most passenger planes
   (b) Orbiting satellites
   (c) Burning meteors
   (d) The Aurora lights

Explain

4. Explain the difference between weather and climate.
5. Draw three diagrams to help you explain the factors that influence the following weather conditions:
   (a) air temperature
   (b) air movement
   (c) the amount of water in the air.

Discover

6. In a magazine or newspaper or online, find a photograph that shows an example of one type of weather. Paste the picture in the centre of a page and add labels about the impact of that weather on the *environment* (for example, creating puddles) and on what we do (such as the clothes people wear).

Predict

7. Look carefully at the map of Australia's climatic zones in figure 3. Predict which two settlements, or *places*, might be at risk of flood. Make sure you explain why you chose them.
8. Look at the *environment* outside the window.
   (a) What is the weather like? Do you think it matches the climatic zone in which you live? Explain.
   (b) Now check to see your climatic zone, using figure 3. If your answers are different, explain why this may have occurred.

Think

9. Look carefully at the photographs in figure 1.
   (a) Describe the weather event in each photograph.
   (b) How would each weather event affect people's lives?
10. Describe how the weather affected you yesterday.

4.3 SkillBuilder: Reading a weather map

online only

WHAT ARE WEATHER MAPS?

Weather maps, or synoptic charts, show weather conditions over a larger area at any given time. They appear every day in newspapers and on television news. Being able to read a weather map is a useful skill because weather affects our everyday life.

Go online to access:

- a clear step-by-step explanation to help you master the skill
- a model of what you are aiming for
- a checklist of key aspects of the skill
- a series of questions to help you apply the skill and to check your understanding.

FIGURE 1 A typical weather map

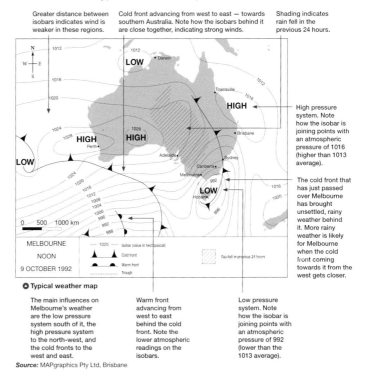

Greater distance between isobars indicates wind is weaker in these regions.

Cold front advancing from west to east — towards southern Australia. Note how the isobars behind it are close together, indicating strong winds.

Shading indicates rain fell in the previous 24 hours.

High pressure system. Note how the isobar is joining points with an atmospheric pressure of 1016 (higher than 1013 average).

The cold front that has just passed over Melbourne has brought unsettled, rainy weather behind it. More rainy weather is likely for Melbourne when the cold front coming towards it from the west gets closer.

MELBOURNE
NOON
9 OCTOBER 1992

1020 — Isobar (value in hectopascal)
Cold front
Warm front
Trough
Rainfall in previous 24 hours

● Typical weather map

The main influences on Melbourne's weather are the low pressure system south of it, the high pressure system to the north-west, and the cold fronts to the west and east.

Warm front advancing from west to east behind the cold front. Note the lower atmospheric readings on the isobars.

Low pressure system. Note how the isobar is joining points with an atmospheric pressure of 992 (lower than the 1013 average).

Source: MAPgraphics Pty Ltd, Brisbane

learn on   RESOURCES — ONLINE ONLY

Watch this eLesson: Reading a weather map (eles-1637)

Try out this interactivity: Reading a weather map (int-3133)

4.4 How is a natural hazard different from a natural disaster?

4.4.1 What are natural hazards?

Australia is prone to a wide variety of **natural hazards**, which range from drought and bushfire to flooding. Many of these events are part of the weather's natural cycle. However, human actions such as overgrazing, deforestation and the alteration of natural waterways have sometimes increased the impact of these hazards. So why do people continue to live in areas that are at risk of experiencing these hazards?

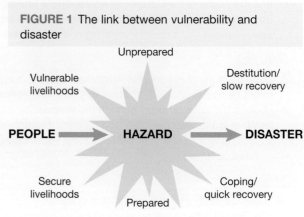

FIGURE 1 The link between vulnerability and disaster

There is a difference between natural hazards and **natural disasters**. A hazard is an event that is a *potential* source of harm to a community. A disaster occurs as the result of a hazardous event that dramatically affects a community. There are four broad types of natural hazard:
1. atmospheric — for example, cyclones, hailstorms, blizzards and bushfires
2. hydrological — for example, flooding, wave action and glaciers
3. geological — for example, earthquakes and volcanoes
4. biological — for example, disease epidemics and plagues.

Clearly natural hazards that are linked to the weather are categorised into the atmospheric and hydrological types. Hazards such as flooding and cyclones could also be termed extreme weather events.

Some natural hazards are influenced by the actions of people and where they choose to locate themselves. For example, the severity of a flood depends not only on the amount and duration of rainfall that occurs. Humans can influence floods by building on floodplains and not planning well for disaster. Environmental degradation and poor urban planning can also turn natural hazards into natural disasters.

Five of Australia's costliest natural disasters
- *Flood*. Queensland, New South Wales and Victoria 2010–2011: 35 deaths, 20 000 homes destroyed in Brisbane alone, $5.6 billion cost
- *Cyclone*. Cyclone Tracy, Darwin 1974: 65 deaths, 10 800 buildings destroyed, $4.18 billion cost
- *Earthquake*. Newcastle 1989: 13 deaths, 50 000 buildings damaged, more than $4 billion cost
- *Hailstorm*. Sydney 1999: 1 death, 24 800 buildings damaged, $2 billion cost
- *Bushfire*. Black Saturday, Victoria 2009: 173 deaths, 3500 buildings destroyed, $1.5 billion cost

Top five casualty rate natural disasters worldwide in the last 15 years
- *Earthquake*. Haiti, 2010: estimated range 85 000–316 000 deaths
- *Tsunami*. Indian Ocean 2004: approximately 230 000 deaths
- *Cyclone*. Cyclone Nargis, Myanmar, 2008: at least 146 000 deaths
- *Earthquake*. Sichuan, China, 2008: approximately 87 400 deaths
- *Earthquake*. Kashmir, Pakistan, 2005: approximately 79 000 deaths

---

4.4.2 Why do people live in areas at risk from natural hazards?

Risk is the possibility of negative effects caused by a natural hazard. Therefore, the type of hazard experienced, along with the **vulnerability** of the people affected, will determine the risk faced. The poorest people in the world are vulnerable because their ability to recover from the impact of a hazard is hampered by their lack of resources. In an event such as a flood or earthquake, people lose their personal belongings, homes and livestock, which are often linked to their incomes, continuing the cycle of poverty. However, in regions that are adequately prepared, and where there is support to cope and rebuild, people recover more quickly.

FIGURE 2 Australia's natural hazards and disasters

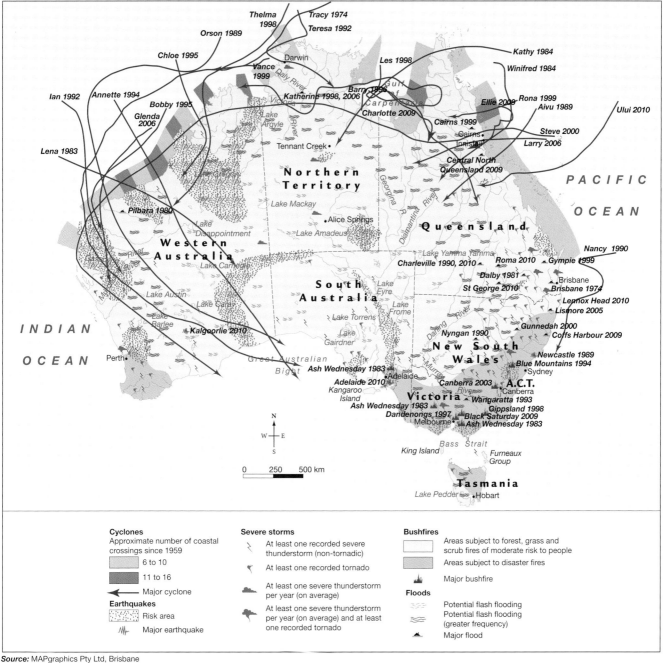

Cyclones
Approximate number of coastal
crossings since 1959
- ▢ 6 to 10
- ▨ 11 to 16
- ← Major cyclone

Earthquakes
- ▒ Risk area
- ⁓ Major earthquake

Severe storms
- ⚡ At least one recorded severe thunderstorm (non-tornadic)
- ⚡ At least one recorded tornado
- ☁ At least one severe thunderstorm per year (on average)
- ☁ At least one severe thunderstorm per year (on average) and at least one recorded tornado

Bushfires
- ▢ Areas subject to forest, grass and scrub fires of moderate risk to people
- ▨ Areas subject to disaster fires
- 🔥 Major bushfire

Floods
- 〰 Potential flash flooding
- 〰 Potential flash flooding (greater frequency)
- ▲ Major flood

Source: MAPgraphics Pty Ltd, Brisbane

4.4 Activities

To answer questions online and to receive **immediate feedback** and **sample responses** for every question, go to your learnON title at www.jacplus.com.au. *Note:* Question numbers may vary slightly.

Remember

1. How do natural hazards and natural disasters differ?

Explain

2. Explain how a flood is both a natural and human hazard.
3. Explain why the risk of experiencing a natural disaster depends on the geographical location of a community.
4. Describe key *changes* that natural hazards and natural disasters can cause to an *environment*.

5. Refer to figure 2.
   (a) What types of natural disasters occur most often in Australia?
   (b) Describe the location of Australia's cyclone hazard zone.
   (c) Give one example of a *place* that has suffered a bushfire disaster.
   (d) What type of hazards are *places* around Newcastle subject to?
   (e) What would be the likely impact of a large earthquake occurring in the earthquake risk area in the Northern Territory?
6. Using the internet, answer the following questions in relation to the flood disaster in Katherine in 1998.
   (a) How did the disaster occur?
   (b) What was the effect of the disaster on the community?
   (c) Why do people risk living there?
7. The casualty rate for the Haitian earthquake is heavily debated. Using the internet, research and suggest possible reasons for a lack of accuracy in this case.

RESOURCES — ONLINE ONLY

Try out this interactivity: Hotspot Commander: natural hazards (int-3083)

4.5 Why does Australia experience droughts?

4.5.1 What is a drought?

Australia is the driest inhabited continent on Earth. As a result, Australia has had many droughts. The main reason Australia is so dry is that much of the continent lies in an area dominated by high atmospheric pressure for most of the year, which brings dry, stable, sinking air to the country. Australia also experiences great variation in its rainfall due to the **southern oscillation** and **El Niño**.

Low average rainfall and extended dry spells are a normal part of life throughout most of Australia. The continent is located in a zone of high pressure that creates conditions of clear skies and low rainfall. Drought conditions occur when the high pressure systems are more extensive than usual, creating long or severe rainfall shortages. A drought is a long period of below-average rainfall, when there is not enough water to supply our normal needs. Because people use water in so many different ways and in such different quantities, there is no universal amount of rainfall that defines a drought.

The term *drought* should not be confused with low rainfall. Sydney could experience a drought and have more rainfall during that period than Alice Springs, which could be experiencing above average rainfall. If low rainfall meant drought, then much of Australia would be in drought most of the time.

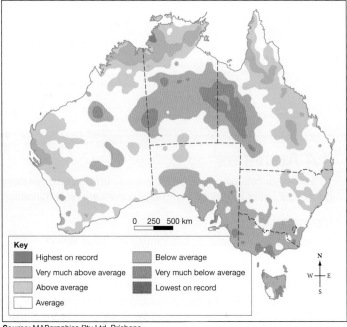

FIGURE 1 Australian rainfall patterns, 2008

Key

Highest on record	Below average
Very much above average	Very much below average
Above average	Lowest on record
Average	

Source: MAPgraphics Pty Ltd, Brisbane

Droughts affect all parts of Australia over a period of time. Intervals between severe droughts have varied from four to 38 years. Some droughts can be localised while other parts of the country receive good rain. Others, such as the drought of 1982–83, can affect more than half the country. Droughts can be short and intense, such as the drought that lasted from April 1982 to February 1983; or they can be long-lived, such as the drought from 2002–2009.

Different weather systems affect different parts of Australia, so there is little chance that all of Australia would be in drought at the same time.

4.5.2 El Niño

Australia's drought of 2002 and beyond, like many before, was caused by what meteorologists call an El Niño event. In a normal year, warm surface water is blown west across the Pacific Ocean towards Australia. This brings heavy rain to northern Australia, Papua New Guinea and Indonesia. On the other side of the Pacific, South America experiences drought. When there is an El Niño event, these winds and surface ocean currents reverse their direction. The warm, moist air is pushed towards South America. This produces rain in South America and drought in Australia.

FIGURE 2 Weather events in (a) a typical year and (b) an El Niño year

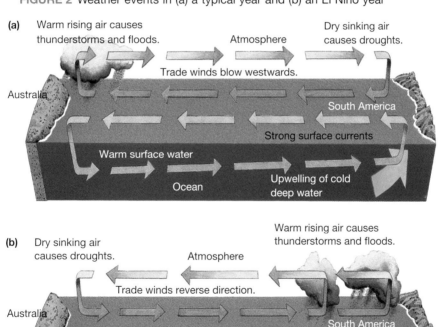

4.5.3 Southern oscillation

Fluctuations in rainfall have several causes that are not fully understood. Probably the main cause of major rainfall fluctuations in Australia is the southern oscillation, which is a major air pressure shift between the Asian and east Pacific regions. The strength and direction of the southern oscillation is measured by a simple index called the southern oscillation index (SOI). The SOI is calculated from monthly or seasonal fluctuations in air pressure between Tahiti and Darwin.

FIGURE 3 Areas affected by El Niño

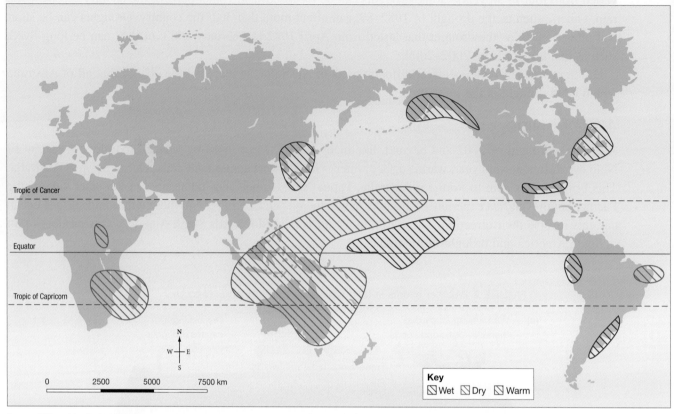

Tropic of Cancer

Equator

Tropic of Capricorn

N
W — E
S

0   2500   5000   7500 km

Key
Wet   Dry   Warm

Source: MAPgraphics Pty Ltd, Brisbane

In an average rainfall year with 'typical' pressure patterns, the SOI is between −10 and +10. If the SOI is strongly negative (below −10), this means that the air pressure at sea level in Darwin is higher than in Tahiti, and an El Niño event occurs.

During an El Niño event, there is less than average rainfall over much of Australia. During this period, drought will occur. If the SOI becomes strongly positive (above +10), this means that the air pressure in Darwin is much lower than normal and a La Niña event occurs. During this period, above average rainfall will occur.

In recent years, scientists have made great advances in understanding and forecasting El Niño and southern oscillation events. The National Climate Centre in Australia produces outlooks on rainfall three months ahead. These outlooks are proving to be of great value to farmers and especially valuable for ecologically sustainable development in rural areas.

4.5 Activities

To answer questions online and to receive **immediate feedback** and **sample responses** for every question, go to your learnON title at www.jacplus.com.au. *Note:* Question numbers may vary slightly.

Remember

1. What is drought?
2. Refer to figure 2, which compares conditions during a typical year and an El Niño year, and study the text. Use the following words to complete the sentences below: stable, moist, cooler, east, drought, Tahiti, dry, warm, north, Darwin.

   During an El Niño event, the normally _____ sea in the oceans to the _____ and _____ of Australia are replaced by much _____ water. The air pressure in _____ begins to fall relative to the air pressure in _____. The normal _____ easterly trade winds change their direction. The result is _____ and _____ air and severe_____.

Explain

3. Why is Australia so dry?
4. What is the SOI and how is it calculated?
5. What do the following SOIs indicate?
   (a) Between +10 and −10
   (b) >+10
   (c) <−10
6. Why is there little chance that all of Australia would be affected by drought at the same time?
7. Refer to figure 3 showing the areas affected by El Niño. Describe the areas that become (a) wetter, (b) drier and (c) warmer during an El Niño event.

Think

8. List some of the short-term effects that drought can have on Australia. Once you have listed these, try to come up with some long-term impacts that Australia and its people would experience if these short-term impacts continued for up to 10 years.
9. Why is El Niño the result of the *interconnection* that occurs between Australia and South America?

RESOURCES — ONLINE ONLY

Watch this eLesson: Weather events in a typical year and an El Niño year (eles-2275)

Try out this interactivity: El Niño (int-3084)

4.6 What are the impacts of drought?

4.6.1 Beyond reasonable drought: Australia, 2002–2009

Around the world, droughts often develop due to similar weather conditions. However, their impacts on different communities that live around the world can be diverse. Often, developed countries find themselves economically worse off during a drought period. In contrast, developing countries usually face devastating social consequences as a result of an extremely dry weather period.

Droughts can last for many years. They may be widespread or confined to small areas. The drought that started in 2002 affected large areas of Australia by 2005 (see figure 1). When Australia experiences a drought, agriculture suffers first and most severely, but eventually everyone feels the impact.

Due to the severe lack of water caused by the drought, many farmers faced production losses because they were not able to sustain their crops or sufficiently

FIGURE 1 Extent of the drought — changes in vegetation cover, 2005

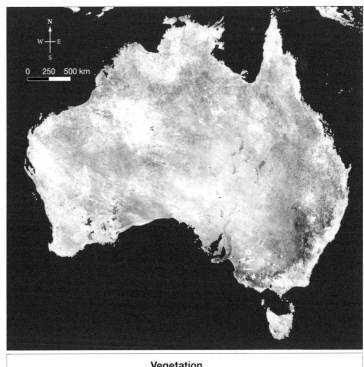

Source: Spatial Vision

feed their livestock. This had negative **economic** impacts.

- By 2004, dairy farmers had experienced a 4.5 per cent drop in their incomes.
- Cotton crops were devastated by the shortage of water.
- Up to 20 cotton communities and approximately 10 000 people in the industry were affected.
- Some communities had to cut production by 60 to 100 per cent.
- Cattle and sheep farmers found it hard to find stockfeed, and prices increased. As a result, herds grew smaller.

In rural towns, jobs were lost and many businesses failed. Some people found themselves forced to leave drought-affected areas in search of other work. Many never returned. Very long droughts cause country people much heartache, and this can result in the break-up of families. It can also lead to severe depression in some individuals. However, the Australian Government set up a fund that farmers and people in agri-cultural businesses can apply to for financial relief when their incomes are disrupted by drought. Counselling hotlines are also available to offer support.

Along with these economic and social impacts, the Australian environment suffers in drought. Droughts have a bad effect on topsoil in Australia. During drought conditions, millions of tonnes of topsoil are blown away (see figure 2). This loss takes many years to replace naturally, if it is ever replaced. The loss of topsoil can make many regions far less productive, making it harder for farmers to recover once the drought has broken.

Use the **News report: dust storms** weblink in the Resources tab to watch a news report on dust storms in Sydney and Brisbane.

4.6.2 The Horn of Africa, 2011

Drought in the Horn of Africa (see figure 3) is becoming a recurring event. Unfortunately, such droughts are occurring in shorter and shorter cycles. As a result, the people who live in this region do not have time to recover before they are faced with another dry period. Recently, the Horn of Africa experienced

FIGURE 2 'Dust' (topsoil) blown from drought-affected inland Australia blankets Sydney, 23 September 2009.

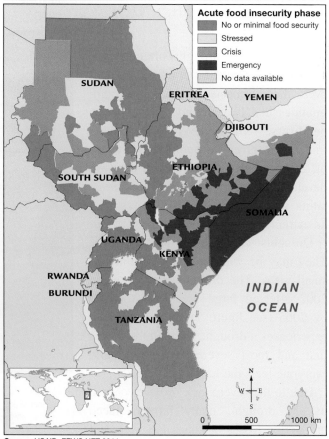

FIGURE 3 Food insecurity in the Horn of Africa, September 2011

Acute food insecurity phase
- No or minimal food security
- Stressed
- Crisis
- Emergency
- No data available

Source: USAID, FEWS NET 2011

drought in 2005–2006, 2009 and most recently 2011–2013. The last disaster is said to be the worst drought in 60 years. These droughts have had terrible social effects.

It is believed that over 10 million people were affected by the drought in 2011. Most people in Ethiopia, Kenya and Somalia are **subsistence farmers**, and rely on agriculture to sustain them. With crop failure and animal losses, a severe food shortage developed, leading to malnutrition. One in three children suffered from malnutrition. The potential for loss of life on a massive scale is a distinct possibility. Tens of thousands are believed to have died in Somalia where **famine** was declared.

The situation in Somalia is made worse by continuing civil unrest. Many of the starving are fleeing to refugee camps in bordering Ethiopia and Kenya. As many as 1500 refugees arrive at these camps every day, many having walked up to six weeks to get there. Due to the extreme poverty of the region, health services are in short supply, adding to these people's problems.

4.6 Activities

To answer questions online and to receive **immediate feedback** and **sample responses** for every question, go to your learnON title at www.jacplus.com.au. *Note:* Question numbers may vary slightly.

Remember
1. Why aren't the people living in the Horn of Africa able to recover from drought?
2. What makes the drought situation worse in Somalia?
3. Identify three key *changes* drought brings to the *environment*.

Explain
4. Why are some drought outcomes in Australia different from those in the Horn of Africa?
5. Describe the *scale* of the 2002–2009 drought.

Think
6. List all the *environmental*, economic and social impacts of drought. Using this list, create a flow diagram to illustrate how these three impact groups relate to, connect to and influence each other. Use this flow diagram to get you started. You can add more boxes and arrows to show how elements are connected.

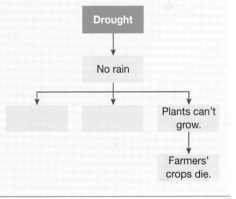

 learn  RESOURCES — ONLINE ONLY

🔗 **Explore more with this weblink:** News report: dust storms

4.7 Why does Australia experience bushfires?

Access this subtopic at **www.jacplus.com.au**

4.8 What happened on Black Saturday?

Access this subtopic at **www.jacplus.com.au**

4.9 How can dry periods be managed to reduce the impact of drought?

4.9.1 Options for managing dry periods

During times of extreme water shortages, governments, communities and individuals often attempt to ensure there is a reliable water supply. The 2002–2009 drought in Australia sparked many different water-saving actions. However, in an environment prone to drought, with increasing demand for water, it is vital to protect and manage water resources at all times — not only during dry periods.

4.9.2 Option 1: government action

The Queensland Government developed the South-East Queensland (SEQ) water grid in order to secure alternative sources of water in an environment that seemed to be growing drier. This strategy aimed to connect the water sources of the region through a pipe network that could move water to different areas and thus meet the needs of local communities. The grid includes existing dams, three water treatment plants and a desalination plant, all connected by approximately 450 kilometres of pipes.

In 2008, the Western Corridor Recycled Project was completed at a cost of $2.5 billion. This project is part of the SEQ water grid and is the largest recycled water scheme in Australia. The project will supply up to 230 megalitres per day of recycled water to industry and power plants. The water also has the potential to be used by farmers and to top up drinking supplies. However, these last two uses of recycled water have created wide debate among communities.

The desalination plant at Tugun on the Gold Coast can provide up to 133 megalitres of drinking water per day. Essentially this project produces drinking water by removing salts and other minerals from sea water. This technology is very successful and has been used in other regions of Australia for years, including Perth, Coober Pedy, Hamilton Island and the Torres Strait islands. Internationally, there are approximately 7500 plants in operation. These desalination plants enable safe drinking water to be produced without having to rely on rainfall.

FIGURE 1 The desalination plant at Tugun produces drinking water for south-east Queensland.

Water measurements
- 1 ML = 1 megalitre
- 1 megalitre = 1 000 000 litres
- 10 megalitres = 4 Olympic-size swimming pools

Stages of the desalination process
1. Sea water is piped from the ocean into a submerged inlet tunnel to the plant.
2. At the pre-treatment stage, particles in the sea water are micro-filtered, the pH is adjusted, and an inhibitor is added to control the build-up of scale in pipelines and tanks.
3. The sea water is forced through layers of membrane to remove salt and minerals. Concentrated salt water is separated and returned to the ocean.
4. During post-treatment, small amounts of lime and carbon dioxide are added to the water, along with chlorine for disinfection.
5. The desalinated water is blended with other Gold Coast water supplies and joins south-east Queensland's water grid to supply homes and industry.

Based on information from www.watersecure.com.au

FIGURE 2 The SEQ water grid

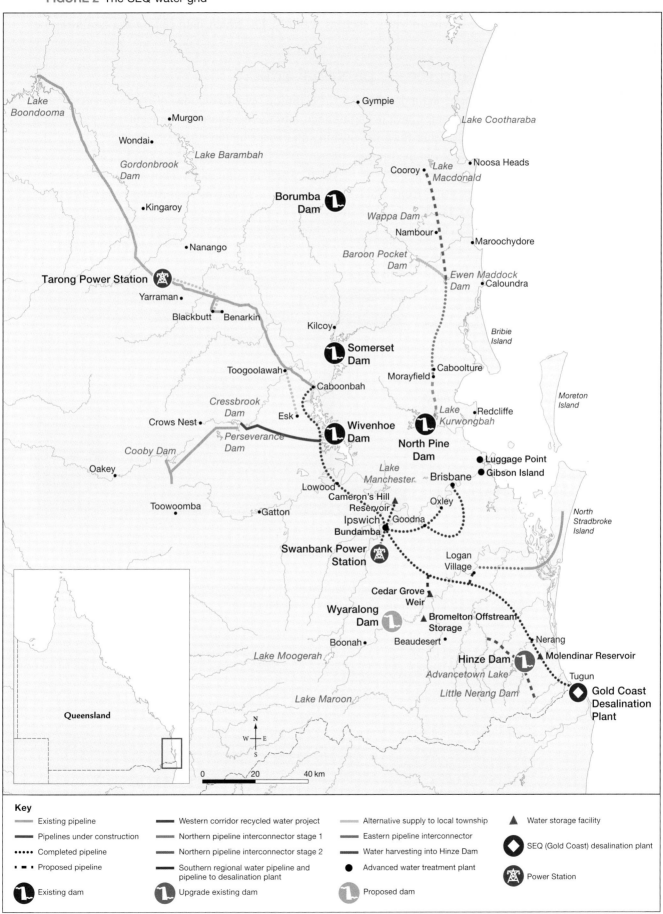

Key

Existing pipeline	Western corridor recycled water project	Alternative supply to local township
Pipelines under construction	Northern pipeline interconnector stage 1	Eastern pipeline interconnector
••••• Completed pipeline	Northern pipeline interconnector stage 2	Water harvesting into Hinze Dam
▪ ▪ ▪ Proposed pipeline	Southern regional water pipeline and pipeline to desalination plant	● Advanced water treatment plant
Existing dam	Upgrade existing dam	Proposed dam

▲ Water storage facility

◆ SEQ (Gold Coast) desalination plant

Power Station

In times of drought, governments may introduce water restrictions to limit the pressure placed on water supplies by individual households and businesses. They may also introduce **incentive** schemes that provide a **rebate** on water-saving devices, such as water tanks, which help relieve the strain on the water supply.

4.9.3 Option 2: You and me — personal action

Owing to government action, more and more people are becoming open to the idea of using recycled water in their homes. In Adelaide, 500 homes have been plumbed into the Southern Urban Reuse Project, which allows them to use recycled waste water for toilet flushing and watering their gardens. This project has the capacity to supply up to 8000 homes in the future.

The global leader in the use of reclaimed, or recycled, water is Singapore. This small island nation used to obtain 50 per cent of its water from Malaysia. However, its goal is to be 100 per cent water independent by 2061. Singapore treats its waste water to a drinkable standard. This method is gaining public acceptance, and will supply up to 50 per cent of the nation's water in future.

Many of our day-to-day actions require us to use water. There are ways we can all use this water more effectively to ensure it is not wasted. By looking at our actions carefully, we can save water in the kitchen, laundry, bathroom and garden (figure 4). Some ideas may include:

- putting aerators on taps
- using a hose with a shut-off nozzle
- cleaning driveways and paths with a broom rather than a hose.

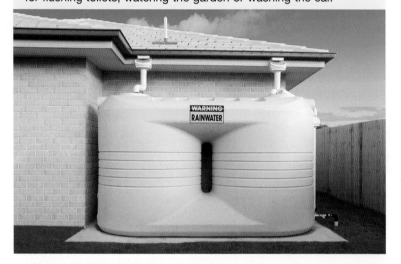

FIGURE 3 Collecting rainwater from your roof can provide water for flushing toilets, watering the garden or washing the car.

4.9 Activities

To answer questions online and to receive **immediate feedback** and **sample responses** for every question, go to your learnON title at www.jacplus.com.au. *Note:* Question numbers may vary slightly.

Remember

1. What is the SEQ water grid made up of?
2. How many litres are there in 230 megalitres?

Explain

3. What is the aim of a desalination plant? How does it achieve this?
4. Will the SEQ water grid be effective in managing water during a drought period? Why or why not?

Think

5. Why do you think the topic of using recycled water can create debate in the community? Create a list of pros and cons for the use of recycled water.
6. Talk to your family about saving water as individuals or as a household. Come up with a list of other ideas that could save water around the home. With this information, create a poster that could be used to educate others.
7. Describe what you believe are the two most *sustainable* ways of reducing the impacts of drought. Give reasons for your choices.

FIGURE 4 Personal action

Ensure your next washing machine has lots of water-efficiency stars.

Ensure you completely fill your dishwasher before using it.

Don't keep the tap running when washing fruit and vegetables. Wash them in a bowl instead.

Install a dual-flush toilet.

Dispose of tissues in the bin — don't flush them down the toilet.

Have short showers. Try for four minutes!

Don't run the tap when brushing your teeth.

Use a water-saving showerhead and keep a bucket in the shower for excess water to use on the garden.

Cover soil in mulch to retain moisture in soil. Grow drought-tolerant plants.

Water the garden in the early morning or evening to reduce evaporation.

4.10 Why does it flood?

4.10.1 Types of floods

Even though Australia is the driest of all the world's inhabited continents, there are periods of very heavy rainfall and flood. Flood disasters in Australia damage property, kill livestock and cause the loss of human life. Since 1788 more than 2000 people have died in floods, equalling the number of deaths from cyclones. In some cases, entire sections of a town have been washed away, as in 1852, when one-third of the town of Gundagai disappeared.

There are three main types of flood:

1. *Slow-onset floods.* These occur along the floodplains of inland rivers, such as the Darling and Namoi, and may last for weeks or months. They are caused by heavy rain and run-off upstream. The water can take days or weeks to affect farms and towns downstream.

2. *Rapid-onset floods.* These occur in mountain headwaters of larger inland rivers or rivers flowing to the coast. The rivers are steeper and the water flows more rapidly. Rapid-onset floods are often more damaging because there is less time to prepare.

3. *Flash floods.* These are caused by heavy rainfall that does not last long, as occurs in a severe thunderstorm. This type of flooding causes the greatest risk to property and human life because it can happen so quickly. It can be a serious problem in urban areas where drainage systems are inadequate.

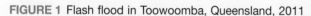

FIGURE 1 Flash flood in Toowoomba, Queensland, 2011

FIGURE 2 Damage left in Toowoomba after the flash flood

4.10.2 Floods and floodplains

Floods are a natural occurrence, but they are a natural hazard to humans, who tend to build farms, towns and transport routes in areas such as floodplains. A floodplain (figure 3) is an area of relatively flat land that borders a river and is covered by water during a flood. Floodplains are formed when the water in a river slows down in flat areas. The river begins to meander and gradually deposits **alluvium**, which builds up the floodplain and other landforms such as deltas.

These fertile, flat areas are used for farming and settlement around the world. In Australia, many of our richest farmlands are on floodplains, and towns are often built on them, close to rivers. Such towns are subject to flooding. The possibility of flood is also increased when vegetation in **catchment areas** has been cleared or modified. Native vegetation can slow down run-off and reduce the chance of flooding.

FIGURE 3 Flat, fertile floodplains are often preferred areas for settlement and farming.

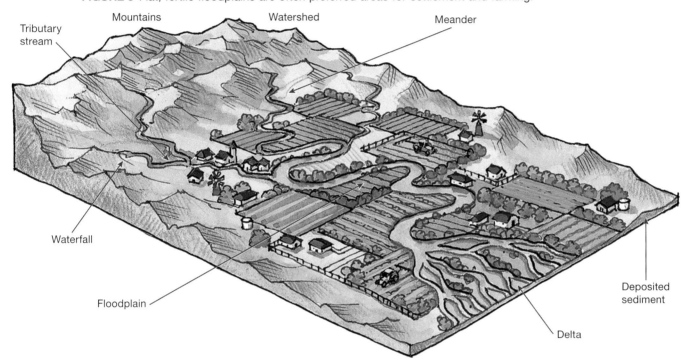

4.10.3 La Niña and floods

A La Niña event in Australia is often associated with floods. La Niña is virtually the opposite of El Niño. Very cold waters dominate the eastern Pacific, and the oceans off Australia are warmer than normal. Large areas of low pressure extend over much of Australia; warm, moist air moves in, and above-average rainfall occurs. There can also be torrential rain and widespread floods.

Recent La Niña events in Australia occurred in 2010–2011, when many parts of Queensland, New South Wales and Victoria were flooded (figure 4).

FIGURE 4 Rockhampton, Queensland, (a) before and (b; on the next page) during the 2011 flood

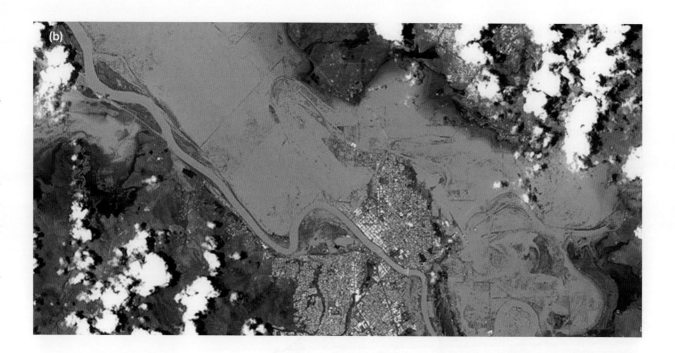

(b)

Remember

1. What are the three main types of floods?
2. What is a floodplain?

Explain

3. Why is alluvium important to agriculture?
4. Construct a simple flow diagram to show the process of a La Niña event (see section 4.10.2).

Predict

5. Should people continue to build on floodplains? Why or why not? Think globally when formulating your argument. Consider *environmental*, cultural and economic factors that could impact on a person's reasoning when choosing a *place* to settle.
6. Create a poster that will warn people about the consequences if they continue to develop on floodplain regions.

Think

7. Why do floods occur on floodplains and in deltas?
8. Create an overlay map to show the distribution of water in the aerial photos of Rockhampton before and after the 2011 flood. Estimate the percentage of the area covered in both maps. (See the SkillBuilder 'Creating an overlay map'.) Describe the *changes* to the *environment*.
9. How might the effects of floods in urban *spaces* differ from effects in rural *spaces*?

4.11 What are the impacts of floods?

4.11.1 The Brisbane floods, 2011

When the Brisbane River broke its banks on 11 January 2011, Australians were shocked and saddened by the devastation left in its wake. Thankfully those affected were able to gain some comfort from the assistance they received from the community as they began the slow process of recovery. However, this is not always an option for those affected by floods in other regions of the world. On the same day on the other side of the world, Brazil was also experiencing a flood. This flood would have a tragic human cost.

Queensland, Australia 2010–2011

Country background

Australia is considered a developed nation with a strong economy. Australians earn on average $38 200 per person. Approximately 24 million people reside in Australia, with 4.8 million of those living in Queensland. About 84 per cent of all Australians are located within 50 kilometres of the coast.

Why?

The flooding that affected this region was due to a strong La Niña event. Long periods of heavy rain over Queensland catchments caused rivers to burst their banks.

Effects

- Three-quarters of the state was declared a disaster zone.
- At least 70 towns and over 200 000 people were affected.
- There were 35 deaths.
- It cost the Australian economy at least $10 billion.
- Up to 300 roads were closed, including nine major highways.
- Over 20 000 homes were flooded in Brisbane alone.
- There was massive damage and loss of property.

Assistance and recovery

- $1.725 billion is being raised by the Federal Government via a flood levy in the tax system.
- $281.5 million is in the Disaster Relief Appeal set up by the then Premier, Anna Bligh.
- Over $20 million was donated to aid agencies such as the St Vincent de Paul Society to help those suffering.
- About $1.2 million was raised through charity sporting events such as Rally for Relief, Legends of Origin and Twenty 20 cricket.
- The Australian Defence Force was mobilised to help with the clean-up.
- The Mud Army was formed: 55 000 volunteers registered to help clean up the streets, and thousands more unregistered people joined them.
- Improvements will be made to dam manuals to help manage the release of water from dams during floods.

FIGURE 1 Anatomy of a flood

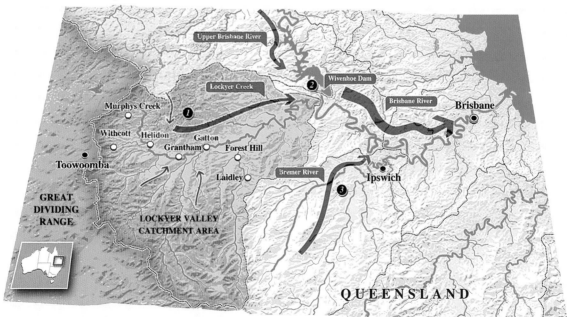

1. Floodwaters from **Lockyer Creek**, which flows into **Brisbane River**. The Lockyer Valley was hit by more than 200 mm of rain.
2. More than 490 000 million litres were released from **Wivenhoe Dam** into Brisbane River.
3. Floodwaters from the **Bremer River**, which is also fed by the **Lockyer Valley**. After passing Ipswich, where it burst its banks, the Bremer River flows into the Brisbane River.

Town heights above sea level in metres: Toowoomba 700 m Murphys Creek 704 m Withcott 262 m
Helidon 143 m Grantham 110 m Gatton 111 m Forest Hill 95 m Laidley 135 m Ipswich 54.8 m Brisbane 28.4 m

CASE STUDY

State of Rio de Janeiro, Brazil, 2011

Country background

Brazil is considered a developing nation. Brazilians earn on average $10 200 per person per year. Approximately 209 million people reside in Brazil, with 650 000 living in the three towns worst affected by the flooding.

Why?

Due to the equivalent of a month's rain falling in 24 hours, flash flooding occurred in a mountainous region in Rio de Janeiro State and São Paulo State. Hillsides and riverbanks collapsed due to landslides. It is believed that illegal construction and deforestation may have contributed to the instability of the land.

Effects

- Approximately 900 people died — most of them in poverty-stricken areas with poor housing conditions and no building policies.
- Forty per cent of the vegetable supply for the city of Rio de Janeiro was destroyed.
- Around 17 000 people were left homeless.
- There was widespread property damage, most of it to homes built riskily at the base of steep hills.

Assistance and recovery

- $460 million was set aside by the President for emergency aid and reconstruction.
- Troops were deployed to help.
- There were donations of clothes and food to the area from other Brazilians.
- About $450 million was loaned by the World Bank.
- Support was given by internal and international charities.

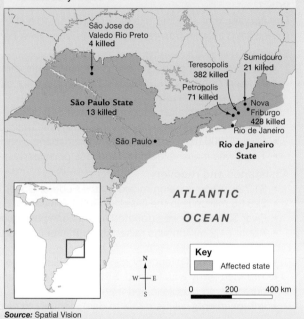

FIGURE 2 Areas affected by the floods in Brazil, 13 January 2011

São Jose do Valedo Rio Preto 4 killed

Teresopolis 382 killed

Sumidouro 21 killed

São Paulo State 13 killed

Petropolis 71 killed

Nova Friburgo 428 killed

Rio de Janeiro

São Paulo

Rio de Janeiro State

ATLANTIC OCEAN

Key

Affected state

0   200   400 km

Source: Spatial Vision

FIGURE 3 Hills collapsed after the heavy rains, destroying homes.

4.11 Activities

To answer questions online and to receive **immediate feedback** and **sample responses** for every question, go to your learnON title at www.jacplus.com.au. *Note:* Question numbers may vary slightly.

Remember

1. When did the floods in Queensland and Brazil occur?

Explain

2. (a) Explain, in your own words, the three causes of the Brisbane floods.
   (b) Explain the causes of the Brazilian floods.
   (c) Compare the causes of these two floods by identifying the similarities and differences between them.
3. (a) Compare the *scale* of these floods.
   (b) Give reasons for the differences in the *scale* of these floods.
4. In a table, classify (in point form) the impacts on people, the economy and the *environment* of the Brisbane and Brazilian floods.
5. Why do you think the impacts differed?
6. How were the responses to the floods in Brisbane and Brazil similar? How and why do you think they were different?

Predict

7. Suggest a few strategies that could be implemented to lessen the impact of floods if they occurred in these regions again.
8. Imagine both these *places* had been given warning that these floods were going to occur. Suggest at least two changes you would expect in relation to the impacts of these two flood events.

Think

9. Imagine you had to evacuate your home because it was under threat from a flood. What five things would you take with you, and why?
10. Describe the *interconnection* between Brisbane and the areas to its west in relation to the January 2011 Brisbane flood.

4.12 How do different places manage floods?

4.12.1 Flood management

Floods occur in many countries around the world. It is important for these countries to learn how to effectively live with this natural hazard. Managing the effects of floods is important if the amount of damage caused is to be minimised. Unfortunately, not all countries have the same resources to tackle this problem. Those countries that are able to invest in flood-prevention **infrastructure** have a greater chance of reducing the risk of flood.

The most common form of flood management is to build a barrier that prevents excess water from reaching areas that would suffer major damage. Levees (see figure 2), **weirs** and dams are a few examples of structures that are built to contain floodwaters. Dams that are used to stop flooding need to be kept below a certain level to allow space for floodwater to fill. Wivenhoe Dam in south-east Queensland was built in response to the floods in 1974 (see figure 1). However, there was some debate about whether this dam could have been used more effectively during the 2010–2011 floods.

FIGURE 1 Lake Wivenhoe, Queensland, at 190 per cent capacity, January 2011

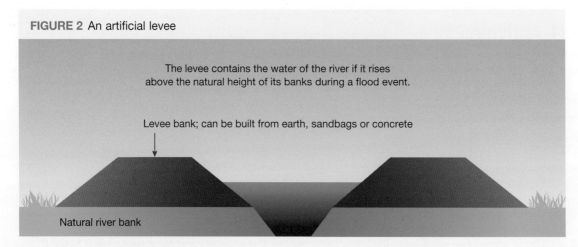

FIGURE 2 An artificial levee

The levee contains the water of the river if it rises
above the natural height of its banks during a flood event.

Levee bank; can be built from earth, sandbags or concrete

Natural river bank

To prevent London being flooded during unusually high tides and storm surges, the city constructed the Thames Barrier, a system of floodgates that stretch across the width of the river (see figure 3). The barrier is triggered if predicted water levels are above a certain height. If this happens, the gates rise to stop the incoming water. Once the water recedes, the danger has passed and the gates are lowered.

FIGURE 3 The Thames flood barrier seen here with its gates up

Another way to manage the risk of damage from floods is to stop building on low-lying land that is guaranteed to flood. Unfortunately, in many urban areas this land has been developed, which increases the chance of property damage in a flood.

Since 2006, the Brisbane City Council has offered a residential property buy-back scheme. This scheme gives people the opportunity to sell their property to the council if they live in a low-lying area that has a 50 per cent chance of flooding every year. People will not be allowed to build on this land again. For this initiative to be successful, it is essential that the price offered by the council is similar to what the owners would get in a private sale; otherwise there is no incentive to use it.

Unfortunately not all countries have the finances to fund property buy-backs or large-scale barrier building. Bangladesh, for example, experiences annual flooding during the **monsoon** season. In response to this, homes are usually built on raised land above flood levels or on stilts.

In order to prepare the population for the arrival of floods, Bangladesh has developed a flood forecasting and warning system that can be broadcast via newspapers, television, radio, the internet and email. Regrettably, due to the growing population in the capital of Dhaka, building is now occurring on low-lying land that was previously used to store floodwater (see figure 4). As a result, many people are still being affected by flooding. In 1998, 65 per cent of Bangladesh was inundated. Twenty million people needed shelter and food aid for two months.

FIGURE 4 In Dhaka, homes are built on stilts to avoid the floodwaters.

4.12 Activities

To answer questions online and to receive **immediate feedback** and **sample responses** for every question, go to your learnON title at www.jacplus.com.au. *Note:* Question numbers may vary slightly.

Remember

1. What flood management techniques are being used in Brisbane?
2. Where are you at most risk from flood when building?

Explain

3. Look at figure 2 and use it to help write a definition for the term levee.
4. How can an early warning system reduce the risk of a flood disaster?
5. Explain the *interconnection* between population growth and the risk posed by floods.

Discover

6. Use the **Bureau of Meteorology** weblink in the Resources tab to find out more about flood warnings. Prepare an information sheet that could be released to a rural community about to be affected by a major flood event. It should include tips on what to do before, during and after the event. Search for the area you live in and check any flood warnings it has had in the past.

Think

7. What do you think would happen if a dam, built to prevent floods, was already full to capacity and the area received more heavy rainfall? What might be some of the consequences?

 RESOURCES — ONLINE ONLY

 Try out this interactivity: Responding to floods (int-3085)

Explore more with this weblink: Bureau of Meteorology

4.13 SkillBuilder: Interpreting diagrams

WHAT ARE DIAGRAMS?

A diagram is a graphic representation of something. In geography, it is often a simple way of showing the arrangement of elements in a landscape and the relationships between those elements. Diagrams also have annotations: labels that explain aspects of the illustration.

Go online to access:

- a clear step-by-step explanation to help you master the skill
- a model of what you are aiming for
- a checklist of key aspects of the skill
- a series of questions to help you apply the skill and to check your understanding.

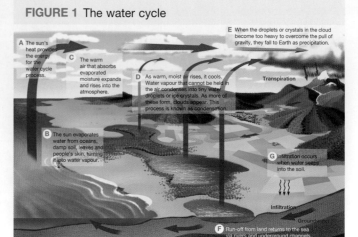

FIGURE 1 The water cycle

A The sun's heat provides the energy for the water cycle process.

B The sun evaporates water from oceans, damp soil, leaves and people's skin, turning it into water vapour.

C The warm air that absorbs evaporated moisture expands and rises into the atmosphere.

D As warm, moist air rises, it cools. Water vapour that cannot be held in the air condenses into tiny water droplets or ice crystals. As more of these form, clouds appear. This process is known as condensation.

E When the droplets or crystals in the cloud become too heavy to overcome the pull of gravity, they fall to Earth as precipitation.

F Run-off from land returns to the sea via rivers and underground channels.

G Infiltration occurs when water seeps into the soil.

Transpiration

Infiltration

Ground

 Watch this eLesson: Interpreting diagrams (eles-1636)

Try out this interactivity: Interpreting diagrams (int-3132)

4.14 Review

4.14.1 Review

The Review section contains a range of different questions and activities to help you revise and recall what you have learned, especially prior to a topic test.

4.14.2 Reflect

The Reflect section provides you with an opportunity to apply and extend your learning.

Access this subtopic at **www.jacplus.com.au**

TOPIC 5
Blow wind, blow

5.1 Overview

Numerous **videos** and **interactivities** are embedded just where you need them, at the point of learning, in your learnON title at www.jacplus.com.au. They will help you to learn the content and concepts covered in this topic.

5.1.1 Introduction

People have long harnessed the power of the wind for energy: we use it to dry clothes, to produce electricity, to pump excess water from the surface of the land and to bring groundwater to the surface.

But strong winds can also cause great destruction, especially when accompanied by heavy rain. These winds can tear roofs from houses and pull trees from the ground.

Starter questions

1. (a) What evidence of wind can you find in the image on this page?
   (b) Outline an impact that wind might have on people and *places*.
2. List as many examples as you can of the way the wind influences you. Include positive and negative influences.
3. As a class, brainstorm a list of extreme weather events related to the wind.
4. Have you ever experienced an extreme weather event? Describe how it made you feel.

A wind farm in a storm: the positive power of the wind

5.2 Why does the wind blow?

5.2.1 Air has weight

Earth's atmosphere protects us from the extremes of the sun's heat and the chill of space, making conditions right to support life. The air in the lowest layer of the atmosphere is called the **troposphere**. Weather is the result of changes in this layer of the atmosphere.

The air around us has weight. The weight of the air above us pushes down on the surface, creating pressure. If we could tie a **barometer** to a hot air balloon, we would see the pressure readings fall as the balloon rose in the atmosphere. This is because there is less air higher up in the atmosphere. You may have read about mountain climbers and athletes having difficulty breathing when they are at high altitudes.

Air pressure

When a person blows up a balloon, the pressure inside the balloon is higher than the surrounding air. When the neck of the balloon is released, the air rushes out of the balloon, as shown in figure 1 (a) and (b). This is wind. If we did not have wind, temperatures would continue to rise over the equator and decrease at the poles.

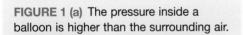

FIGURE 1 (a) The pressure inside a balloon is higher than the surrounding air.

FIGURE 1 (b) When the neck of the balloon is released, air rushes out, moving from a space of high pressure to one of low pressure.

High pressure

Low pressure

Meteorologists are able to measure air pressure using a unit of measure called a millibar. The average weight of air is about 1013 millibars. Measurements higher than this indicate areas of high pressure; here, the air is sinking. Measurements lower than 1013 millibars indicate areas of low pressure; here, the air is rising. Wind is caused by air moving from areas of high pressure to areas of low pressure.

5.2.2 Why does air pressure vary across the Earth?

Variations in air pressure are the result of the heating effect of the sun and the rotation of the Earth.

Effects of the sun

The warming influence of the sun varies with the time of day (see figure 2) and latitude (distance from the equator). Temperatures are higher in the middle of the day, and higher at the equator than at the poles.

Warm air is also less dense than cold air. This is because as the air heats, it expands, causing it to rise. Air pressure over the equator is less than at the poles. As the warm air over the equator rises and expands, cooler air from near the poles rushes in to replace it. As a result, air is circulated around the Earth, and this movement of air is what we call wind.

Effect of the Earth's rotation

The rotation of the Earth on its axis causes the air above the surface of the Earth to be deflected rather than to travel in a straight line. This causes the wind to circle around high and low pressure systems. The direction in which winds circle depends on whether you are in the Northern or Southern Hemisphere. As the air moves from an area of high pressure to an area of low pressure, winds circle in the opposite direction in each hemisphere. In an area of high pressure, the winds circle in an anticlockwise direction in the Southern Hemisphere and a clockwise direction in the Northern Hemisphere. This deflection of winds is known as the Coriolis effect (see figure 3).

FIGURE 2 On a smaller scale, this diagram shows the effect of the sun on a sea breeze.

Land heats up and cools down more quickly than the sea.

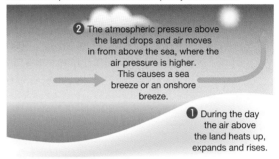

❷ The atmospheric pressure above the land drops and air moves in from above the sea, where the air pressure is higher. This causes a sea breeze or an onshore breeze.

❶ During the day the air above the land heats up, expands and rises.

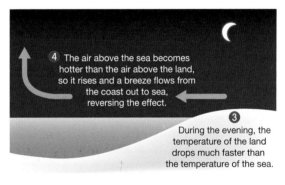

❹ The air above the sea becomes hotter than the air above the land, so it rises and a breeze flows from the coast out to sea, reversing the effect.

❸ During the evening, the temperature of the land drops much faster than the temperature of the sea.

FIGURE 3 Wind is caused by air moving from areas of high pressure to areas of low pressure. Its direction is influenced by the rotation of the Earth.

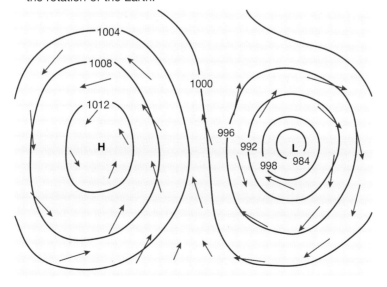

5.2 Activities

To answer questions online and to receive **immediate feedback** and **sample responses** for every question, go to your learnON title at www.jacplus.com.au. *Note:* Question numbers may vary slightly.

Remember

1. What is wind?
2. Identify the two factors that influence wind. Would either of these factors influence the strength of the wind? Explain.

Explain

3. Explain why you are not affected by the pressure of the atmosphere.
4. What role does the sun play in causing wind?

Discover

5. Working with a partner, try the following activity to illustrate the influence of the Earth's rotation on wind.
   Step 1. Place a pin through the centre of a piece of paper and attach it to a piece of cardboard. Make sure the paper can move freely on the pin.
   Step 2. Have your partner rotate the piece of paper around the pin. At the same time, you should attempt to draw a straight line on the page.
   Step 3. Record your findings.
   Step 4. Compare and discuss your results with the class.
6. Over the course of the next week, collect weather maps from the daily newspaper and find your location.
   (a) Is the weather being influenced by a high or a low pressure system?
   (b) Will the wind be moving in a clockwise or anticlockwise direction? Give reasons for your answer.
7. Observe the wind conditions outside your classroom.
   (a) Is it windy today?
   (b) Identify the direction in which the wind is blowing.

Predict

8. How easy is it to predict the weather? People often complain that the forecasters don't always get it right.
   (a) Using the weather maps you collected earlier, and the observations you made, write a weather forecast for tomorrow. In your forecast, make reference to both wind speed and direction.
   (b) Collect tomorrow's weather map and make observations similar to those you made in question 6. Record your findings
   (c) Compare what you have written for this activity. How accurate were your predictions? Suggest factors that might influence the accuracy of such predictions and *changes* that you observe.
9. What is the *interconnection* between our atmosphere and the weather we experience at the Earth's surface?

learn on RESOURCES — ONLINE ONLY

 Try out this interactivity: Highs and lows (int-3086)

my**World**Atlas Deepen your understanding of this topic with related case studies and questions.
❯ **Australia: weather and climate**

5.3 How strong is the wind?

5.3.1 How is wind shown on a weather map?

Differences in air pressure lead to variations in the strength of the wind. You can work out the strength of the wind by looking at weather maps, the behaviour of objects or by using instruments designed to measure the strength of the wind. Winds are named according to their source. This means that a northerly wind is coming from the north and a southerly from the south.

If you study the **isobars** on a weather map you will notice that they are not evenly spaced. Look closely at the map in figure 1 (a). The wind is strongest in the southern regions of this map, where the isobars

FIGURE 1 (a) A typical weather map

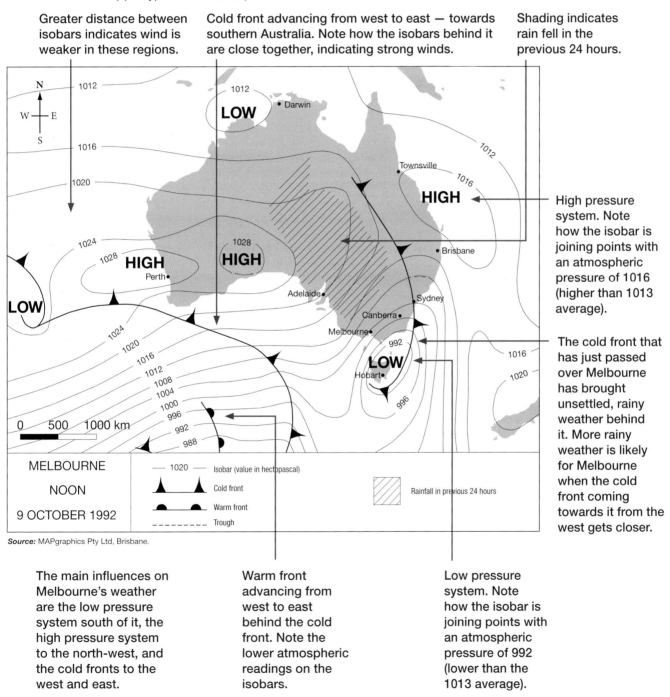

Greater distance between isobars indicates wind is weaker in these regions.

Cold front advancing from west to east — towards southern Australia. Note how the isobars behind it are close together, indicating strong winds.

Shading indicates rain fell in the previous 24 hours.

High pressure system. Note how the isobar is joining points with an atmospheric pressure of 1016 (higher than 1013 average).

The cold front that has just passed over Melbourne has brought unsettled, rainy weather behind it. More rainy weather is likely for Melbourne when the cold front coming towards it from the west gets closer.

MELBOURNE
NOON
9 OCTOBER 1992

1020 — Isobar (value in hectopascal)
Cold front
Warm front
Trough

Rainfall in previous 24 hours

Source: MAPgraphics Pty Ltd, Brisbane.

The main influences on Melbourne's weather are the low pressure system south of it, the high pressure system to the north-west, and the cold fronts to the west and east.

Warm front advancing from west to east behind the cold front. Note the lower atmospheric readings on the isobars.

Low pressure system. Note how the isobar is joining points with an atmospheric pressure of 992 (lower than the 1013 average).

are close together, and gentler in the northern parts of the map, where the spacing between them is much greater. The symbols shown in figure 1 (b) are also commonly used on weather maps to give a more accurate representation of wind speed and to provide information on the direction of the wind.

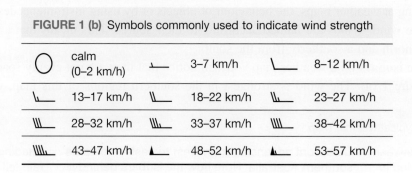

FIGURE 1 (b) Symbols commonly used to indicate wind strength

◯	calm (0–2 km/h)	⌐	3–7 km/h	⌐	8–12 km/h
	13–17 km/h		18–22 km/h		23–27 km/h
	28–32 km/h		33–37 km/h		38–42 km/h
	43–47 km/h		48–52 km/h		53–57 km/h

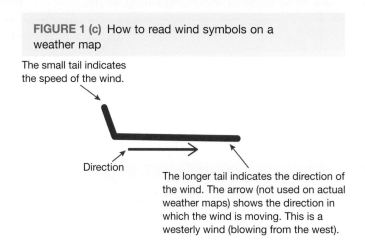

FIGURE 1 (c) How to read wind symbols on a weather map

The small tail indicates the speed of the wind.

Direction

The longer tail indicates the direction of the wind. The arrow (not used on actual weather maps) shows the direction in which the wind is moving. This is a westerly wind (blowing from the west).

Can I make observations about the speed of the wind?

The Beaufort scale (see figure 3) relates wind speed to the observable movement of objects within the environment.

Using a wind rose

A wind rose such as that shown in figure 2 uses data collected over long periods of time to visually represent wind information. The spokes represent wind direction; the longer the spoke the more frequently the wind blows from a particular direction. The thickness of the bands represents the speed of the wind. Refer to the SkillBuilder 'Cardinal points: wind roses' in subtopic 5.4 to learn how to use a wind rose.

FIGURE 2 A wind rose can show wind speed, direction and frequency over a long period of time.

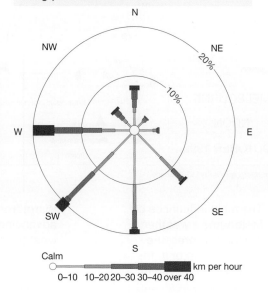

Calm

km per hour
0–10 10–20 20–30 30–40 over 40

FIGURE 3 The Beaufort scale is based on the observable impact of winds.

0 Calm
Less than 2 km/h
Smoke rises vertically

2 Light breeze
6–12 km/h
Wind felt on face,
wind vanes move

4 Moderate breeze
21–30 km/h
Dust and loose paper
move, small branches
move

6 Strong breeze
41–51 km/h
Large branches move,
umbrellas difficult to use,
difficult to walk steadily

8 Gale
64–77 km/h
Twigs broken off trees,
difficult to walk

10 Whole gale
88–101 km/h
Trees uprooted,
considerable
structural damage

12 Hurricane/cyclone
Greater than 120 km/h
Widespread devastation

1 Light air
2–5 km/h
Smoke drift shows wind
direction, wind vanes
don't move

3 Gentle breeze
13–20 km/h
Leaves and small twigs in
motion, hair disturbed,
clothing flaps

5 Fresh breeze
31–40 km/h
Small trees with leaves
begin to sway, wind force
felt on body

7 Moderate gale
52–63 km/h
Whole trees in motion,
inconvenience felt
when walking

9 Strong gale
78–86 km/h
People blown over, slight
structural damage, including
tiles blown off houses

11 Storm
102–120 km/h
Widespread damage

5.3 Activities

To answer questions online and to receive **immediate feedback** and **sample responses** for every question, go to your learnON title at www.jacplus.com.au. *Note:* Question numbers may vary slightly.

Remember

1. What *change* does difference in air pressure cause?
2. Describe two methods you could use to determine wind speed.
3. In your opinion, which of these methods gives the most useful information about wind speed? Give reasons for your answer.

Explain

4. Using figure 1, describe the wind speeds and directions in Western Australia and along the east coast of Australia on that day.

Predict

5. Collect a weather map from the newspaper or online. Based on what you know about weather and reading weather maps predict what the wind conditions will be like in each of the Australian capital cities over the next day or so.

   Use the **Wind rose maps** weblink in the Resources tab to compare your predictions with what is shown on these maps. Make sure you select the current month. Note any similarities and differences. Why might differences occur?

Think

6. Devise your own symbols similar to those shown on the Beaufort scale. Obtain a current weather map from the newspaper or online. Paste it onto a sheet of paper and annotate your map with your symbols for describing wind speed. Swap maps with a partner and further annotate each other's maps with written descriptions of the symbols shown.

myWorldAtlas   **Deepen your understanding of this topic with related case studies and questions.**
◊ Wind and sun direction

5.4 SkillBuilder: Cardinal points: wind roses

WHAT ARE WIND ROSES?

A wind rose is a diagram that shows the main wind features of a place, in particular wind direction, speed and frequency. Wind directions can be divided into eight or 16 compass directions.

Go online to access:

- a clear step-by-step explanation to help you master the skill
- a model of what you are aiming for
- a checklist of key aspects of the skill
- a series of questions to help you apply the skill and to check your understanding.

FIGURE 1 A wind rose

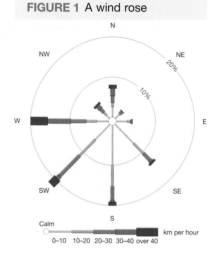

5.5 How do thunderstorms form?

5.5.1 What causes thunderstorms?

Thunderstorms, also referred to as electrical storms, form in unstable, moist atmospheres where powerful updrafts occur, as happens when a cold front approaches. It is estimated that, around the Earth, there are 1800 thunderstorms each day. Over the past 12 years, an average of around 100 severe thunderstorms were reported in Australia each year.

Some 1000 or so years ago, the Vikings thought thunder was the rumble of Thor's chariot. (He was their god of thunder and lightning.) Lightning marked the path of his mighty hammer Mjollnir when he threw it across the sky at his enemies.

Today we know that thunderstorms occur when large **cumulonimbus clouds** build up enough static electricity to produce lightning, as shown in figure 1. Lightning instantly heats the air through which it travels to about 20 000 °C — more than three times as hot as the surface of the sun. This causes the air to expand so quickly that it produces an explosion (thunder). The time between a lightning flash and the crash of thunder tells you how far away the lightning is (5 seconds means that the lightning is 1.6 kilometres away).

FIGURE 1 How a thunderstorm works

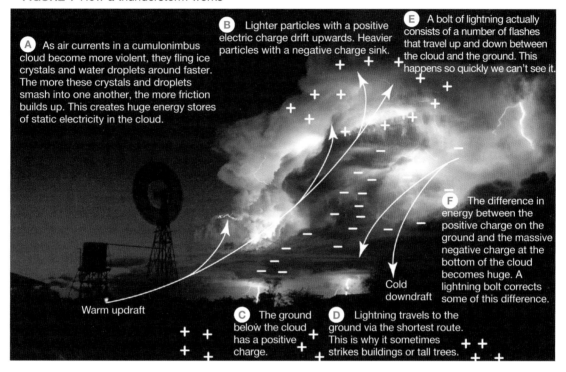

B Lighter particles with a positive electric charge drift upwards. Heavier particles with a negative charge sink.

E A bolt of lightning actually consists of a number of flashes that travel up and down between the cloud and the ground. This happens so quickly we can't see it.

A As air currents in a cumulonimbus cloud become more violent, they fling ice crystals and water droplets around faster. The more these crystals and droplets smash into one another, the more friction builds up. This creates huge energy stores of static electricity in the cloud.

F The difference in energy between the positive charge on the ground and the massive negative charge at the bottom of the cloud becomes huge. A lightning bolt corrects some of this difference.

Cold downdraft

Warm updraft

C The ground below the cloud has a positive charge.

D Lightning travels to the ground via the shortest route. This is why it sometimes strikes buildings or tall trees.

5.5.2 Severe thunderstorms

According to the Bureau of Meteorology, a thunderstorm can be classified as severe if it has one or more of the following features.

- Flash flooding. Thunderstorms often move slowly, dropping a lot of precipitation in one area. The rain or hail may consequently be too heavy and long-lasting for the ground to absorb the moisture. The water then runs off the surface, quickly flooding local areas.
- **Hailstones** that are two centimetres or more in diameter. The largest recorded hailstone had a circumference of 47 centimetres.
- Wind gusts of 90 kilometres per hour or more. Cold blasts of wind hurtle out of thunderclouds, dragged down by falling rain or hail. When the drafts hit the ground, they gust outwards in all directions.

In the right conditions, tornadoes can occur (see subtopic 5.10). These are rapidly spinning updrafts of air that can develop as a result of thunderstorm activity. Although severe tornadoes are not common in Australia, around 400 tornadoes have been recorded.

FIGURE 2 The Vikings believed the god Thor produced lightning and thunder.

5.5.3 When do thunderstorms occur?

Thunderstorms can occur at any time of the year, but they are more likely to occur during spring and summer, as shown in figures 3 and 4. This is due mainly to the warming effects of the sun and the fact that warm air can hold more moisture than cold air.

Thunderstorms are created when cooler air begins to push warmer, humid air upwards. As the warm air continues to rise rapidly in an unstable atmosphere, the cloud builds up higher and begins to spread. Thunderstorms can quickly develop when the atmosphere remains unstable or when it is able to gather additional energy from surrounding winds.

The time of day when thunderstorms are more likely is shown in figure 5. You will notice

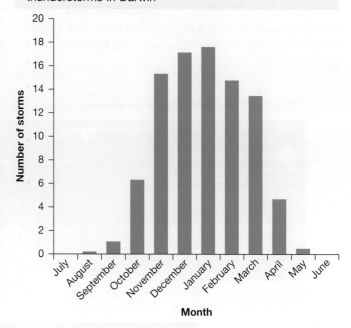

FIGURE 3 Average monthly distribution of thunderstorms in Darwin

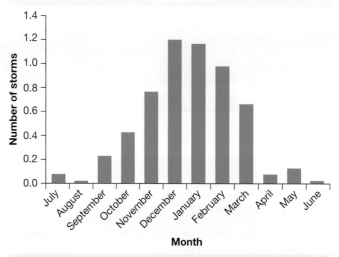

FIGURE 4 Average monthly distribution of thunderstorms in Hobart

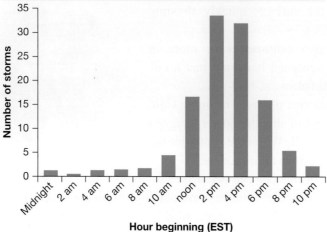

FIGURE 5 Hourly distribution of thunderstorms in New South Wales and the Australian Capital Territory

that thunderstorm activity is greater in the afternoon. This is linked to the daily heating of the Earth by the sun, which peaks in the afternoon.

5.5.4 What is a hailstorm?

When we think about thunderstorms, we often think only of the high winds, thunder and lightning, but significant damage is also caused by hailstones. Any thunderstorm that produces hailstones large enough to reach the ground is known as a **hailstorm**. Hailstones in Australia tend to range in size from a few millimetres to the size of a tennis ball (see figure 7).

5.5.5 Inside a storm

January 2016 saw widespread supercell storm activity across Queensland, New South Wales, Victoria and South Australia.

On 13 January, Melbourne sweltered through temperatures of about 43 °C. Intense thunderstorm activity with wind gusts up to 100 kilometres per hour swept through in the early evening causing the city to be blanketed by a cloud of dust. Up to 1000 homes were left without power.

The following day a severe storm struck Sydney with winds gusting up to 98 kilometres per hour bringing down power lines, damaging buildings and cars and causing flash flooding. More than 40 000 homes and businesses reported power outages. The temperature plummeted by more than 10 °C in five minutes. Emergency services responded to 145 storm-related incidents, including a gas leak.

On 16 January, Townsville recorded 91 millimetres of rain in 30 minutes, resulting in flash flooding leaving many motorists stranded. The rain continued to fall, with 181 millimetres recorded in two hours. Wind gusts of more than 100 kilometres accompanied the massive storm that has been described as a once-in-a-100-year event. Unfortunately, while large areas were inundated, the rain had little impact on the region's water storages.

Both Adelaide and Sydney were pummelled by supercell storms on

FIGURE 6 How a supercell hailstorm forms

As the ice crystals rise and fall with the updrafts, they collide with water droplets that freeze around them, forming hailstones.

The more the hailstones rise and fall, the larger they become.

Hail begins to form when strong updrafts lift moist air into the freezing zone in cumulonimbus clouds. Here, the moisture in the air turns to ice crystals.

Hailstones fall to Earth (at speeds of around 160 km/h) when they become too heavy or are caught in a downdraft.

A supercell storm has extremely strong updrafts. These keep hailstones suspended longer than regular storms, allowing hailstones to grow bigger. To form cricket-ball sized hailstones, there must be updrafts of around 160 km/h.

FIGURE 7 Hailstones can be the size of a golf ball or bigger.

FIGURE 8 The force of a storm tore this tree from the ground.

22 January. The worst hit areas were in the Adelaide Hills and Fleurieu Peninsula where 20 000 homes lost power and the SES responded to 61 calls for help. Thirty-five millimetres of rain was recorded in half an hour, resulting in flash flooding and hailstorms measuring two centimetres in diameter carpeting parts of the city. Wind gusts of up to 90 kilometres per hour were recorded at the airport.

Meanwhile, Sydney was warned to prepare for the worst, to secure vehicles and loose items, unplug electronic equipment and to stay indoors as the city braced itself for more storms, following on from those experienced in previous days. The intense storm activity was the result of the large number of hot days. Flash flooding, damaging winds, hail and lightning were set to continue.

On 29 January, the tourist hot spots around the Gold Coast and Sunshine Coast were lashed by severe storm activity. Wind gusts of more than 100 kilometres per hour were recorded, with almost 9000 properties losing power.

FIGURE 9 In June 2016, another supercell storm hit Sydney. Waves up to eight metres high crashed into the shoreline at Collaroy Beach and caused extensive damage.

5.5.6 How do I protect myself in a thunderstorm?

During storms, damage and injury are often caused by loose objects blown around by the wind, by lightning strikes, and by people being caught in flash floods. To protect yourself:

- before the storm approaches, make sure loose objects outside your home are secure
- stay inside during the storm
- unplug electrical equipment such as computers, televisions and gaming consoles
- avoid using the phone until the storm has passed
- use torches rather than candles as a source of light
- stay indoors, and stay away from windows
- if caught in a storm, try to find shelter
- if caught in the open, move away from objects that could fall, such as trees
- crouch down; don't huddle in a group
- never try to walk or drive through floodwater
- do not touch or approach fallen power lines.

FIGURE 10 The roof of a house sits in the middle of the road at The Gap in Brisbane's north-west. The roof is from a home 50 metres away.

5.5 Activities

To answer questions online and to receive **immediate feedback** and **sample responses** for every question, go to your learnON title at www.jacplus.com.au. *Note:* Question numbers may vary slightly.

Remember

1. What is a thunderstorm?
2. (a) List the *changes* to the *environment* and types of damage that might result from thunderstorm activity.
   (b) Next to each type of damage indicate:
   - whether the damage is caused predominantly by wind or water
   - whether the damage tends to occur to the natural or built *environment*.
3. Study figure 6 and explain how hailstones are formed.

Explain

4. Suggest reasons why people in earlier civilisations assumed weather events were the work of the gods.
5. Explain why thunderstorms can cause so much damage to the natural and human *environments*.
6. Use the diagrams on this page to make your own sketch of a supercell storm. Using words such as evaporation, condensation and precipitation, annotate your diagram to explain how storms develop.

Predict

7. Study figure 6, showing a supercell storm. Write a paragraph explaining why hailstones can vary so much in size.
8. During which seasons of the year are thunderstorms more likely? Give reasons for your answer.

Think

9. Study figure 5. During which hours of the day do most severe thunderstorms occur? Why do you think this is so?
10. (a) Use the information in this spread to annotate a map of Australia to show the dates when thunderstorms were recorded around Australia and the damage they caused. Use the internet to find information to annotate *places* that are not mentioned in this spread.
    (b) Explain why so much thunderstorm activity occurs during January.
11. Select three points from the list of actions on how to protect yourself in a thunderstorm. Explain the rationale between the points you have chosen.
12. Use **The Gap storm** weblink in the Resources tab to watch a video filmed by two men in the Brisbane suburb of The Gap in 2008. After watching the video, describe what you saw. Where would the film-makers have been standing when they shot this footage? Identify safety rules that the film-makers have ignored.

 RESOURCES – ONLINE ONLY

Explore more with this weblink: The Gap storm

5.6 What is a cyclone?

5.6.1 What is a cyclone?

Tropical **cyclones** (called hurricanes in the Americas and **typhoons** in Asia) can cause great damage to property and significant loss of life. Some 80 to 100 tropical cyclones occur around the world every year in tropical coastal areas located north and south of the equator. Australia experiences, on average, about 13 cyclones per year.

Cyclones form when a cold air mass meets a warm, moist air mass lying over a tropical ocean with a surface temperature greater than 27 °C. Cold air currents race in to replace rapidly rising, warm, moist air currents, creating an intense low pressure system. Winds with speeds over 119 kilometres per hour can be generated. Cyclones are classified using the scale in table 1.

Figure 2 shows the continuous cycle of evaporation, condensation and precipitation associated with cyclones. At first the winds spin around an area about 200 to 300 kilometres wide. As the winds gather energy by sucking in more warm moist air, they get faster. In severe cyclones, winds may reach speeds of 295 kilometres per hour. The faster the winds blow, the smaller the area around which they spin; this is called the eye. It might end up being only about 30 kilometres wide. Around the edge of the eye, winds and rain are at their fiercest. However, in the eye itself, the air is relatively still, and the sky above it may be cloudless.

TABLE 1 Classification of cyclones using the Saffir–Simpson scale

Category	Wind gust speed/ ocean swell	Damage
1	Less than 125 km/h 1.2–1.6 m	Mild damage
2	126–169 km/h 1.7–2.5	Significant damage to trees
3	170–224 km/h 2.6–3.7 m	Structural damage, power failures likely
4	225–279 km/h 3.8–5.4 m	Most roofing lost
5	More than 280 km/h More than 5.4 m	Almost total destruction

What damage is caused by tropical cyclones?

Tropical cyclones can cause extensive damage if they cross land. **Gale force winds** can tear roofs off buildings and uproot trees. **Torrential rain** can often cause flooding, as can **storm surges**.

When a tropical cyclone approaches or crosses a coastline, the very low atmospheric pressure and impact of strong winds on the sea surface combine to produce a rise in sea level, as shown in figure 3.

FIGURE 1 World distribution of tropical cyclones by names used in different regions

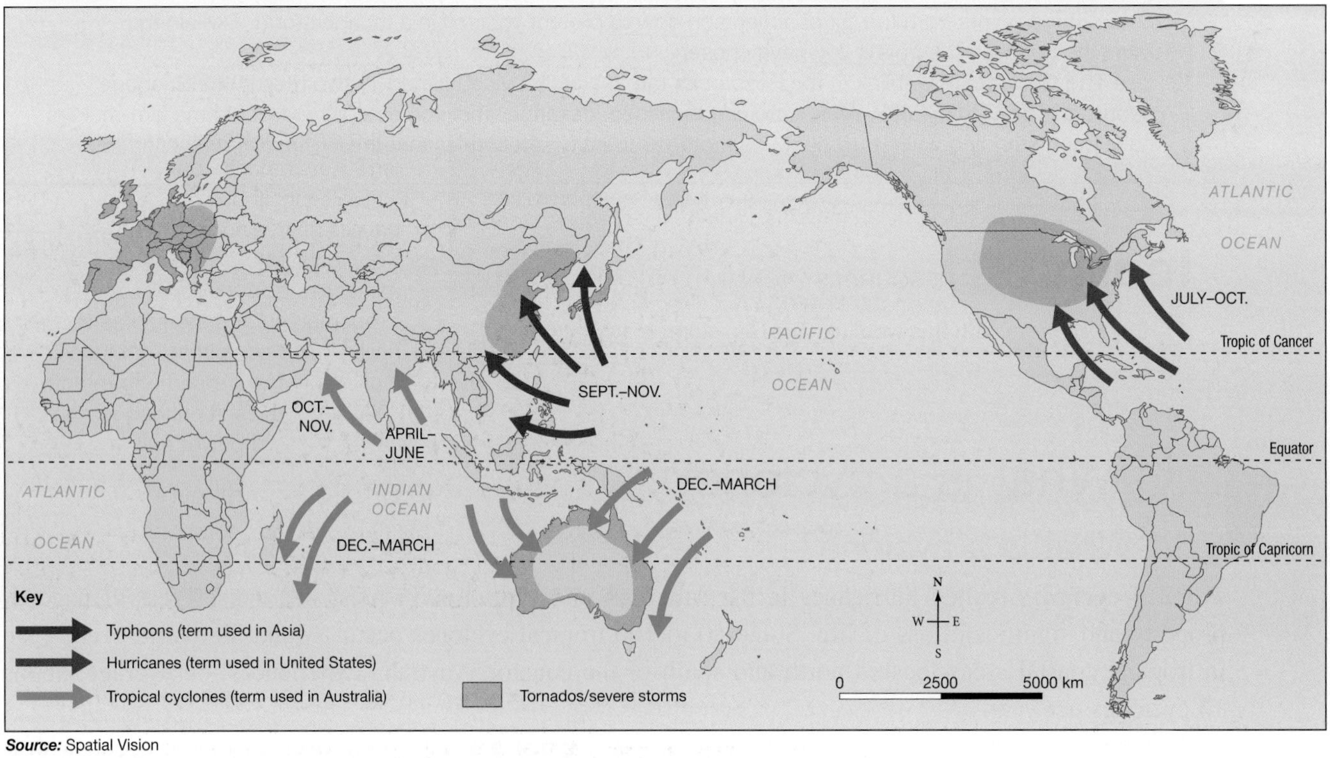

Key

Typhoons (term used in Asia)

Hurricanes (term used in United States)

Tropical cyclones (term used in Australia)

Tornados/severe storms

Source: Spatial Vision

FIGURE 2 How a cyclone forms. The winds within a cyclone spin because of the rotation of the Earth. In the Southern Hemisphere, they rotate in a clockwise direction. In the Northern Hemisphere, they rotate in an anticlockwise direction.

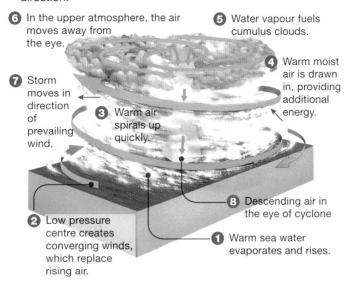

6 In the upper atmosphere, the air moves away from the eye.

5 Water vapour fuels cumulus clouds.

4 Warm moist air is drawn in, providing additional energy.

7 Storm moves in direction of prevailing wind.

3 Warm air spirals up quickly.

8 Descending air in the eye of cyclone

2 Low pressure centre creates converging winds, which replace rising air.

1 Warm sea water evaporates and rises.

FIGURE 4 Satellite image of Hurricane Katrina, which caused massive damage in New Orleans in 2005

FIGURE 3 Flooding caused by storm surges

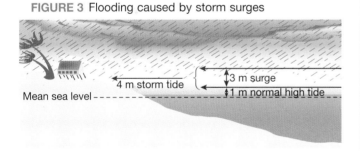

3 m surge
4 m storm tide
1 m normal high tide
Mean sea level

FIGURE 5 The power of a cyclone

5.6.2 Cyclone Winston

CASE STUDY

How did Cyclone Winston impact Fiji?

On 7 February 2016, a tropical disturbance was noted north-west of Port Vila, Vanuatu, tracking in a south-easterly direction. By 11 February it had acquired gale-force winds. Over the next few days Cyclone Winston went through a cycle of intensifying, weakening and stalling until finally developing into a category 5 cyclone on 19 February. The following day, shortly before making landfall on Viti Levu, Fiji, Cyclone Winston reached its peak intensity. Sustained winds of 230 kilometres per hour, with gusts of up to 285 kilometres per hour and a central pressure reading of 915 millibars, were recorded.

How much damage was caused?

Cyclone Winston has been described as the most powerful storm to strike in the Southern Hemisphere. Strong winds battered the island nation of Fiji with damage multiplied by a 4-metre storm surge (see section 5.6.1).

The damage bill has been estimated at more than US$650 million. More than 40 people were killed and communication was cut, leaving at least six outer islands isolated for days. In the years prior to Cyclone Winston, the Fijian Government had invested heavily in infrastructure, much of which was washed away.

FIGURE 6 A track map of Cyclone Winston

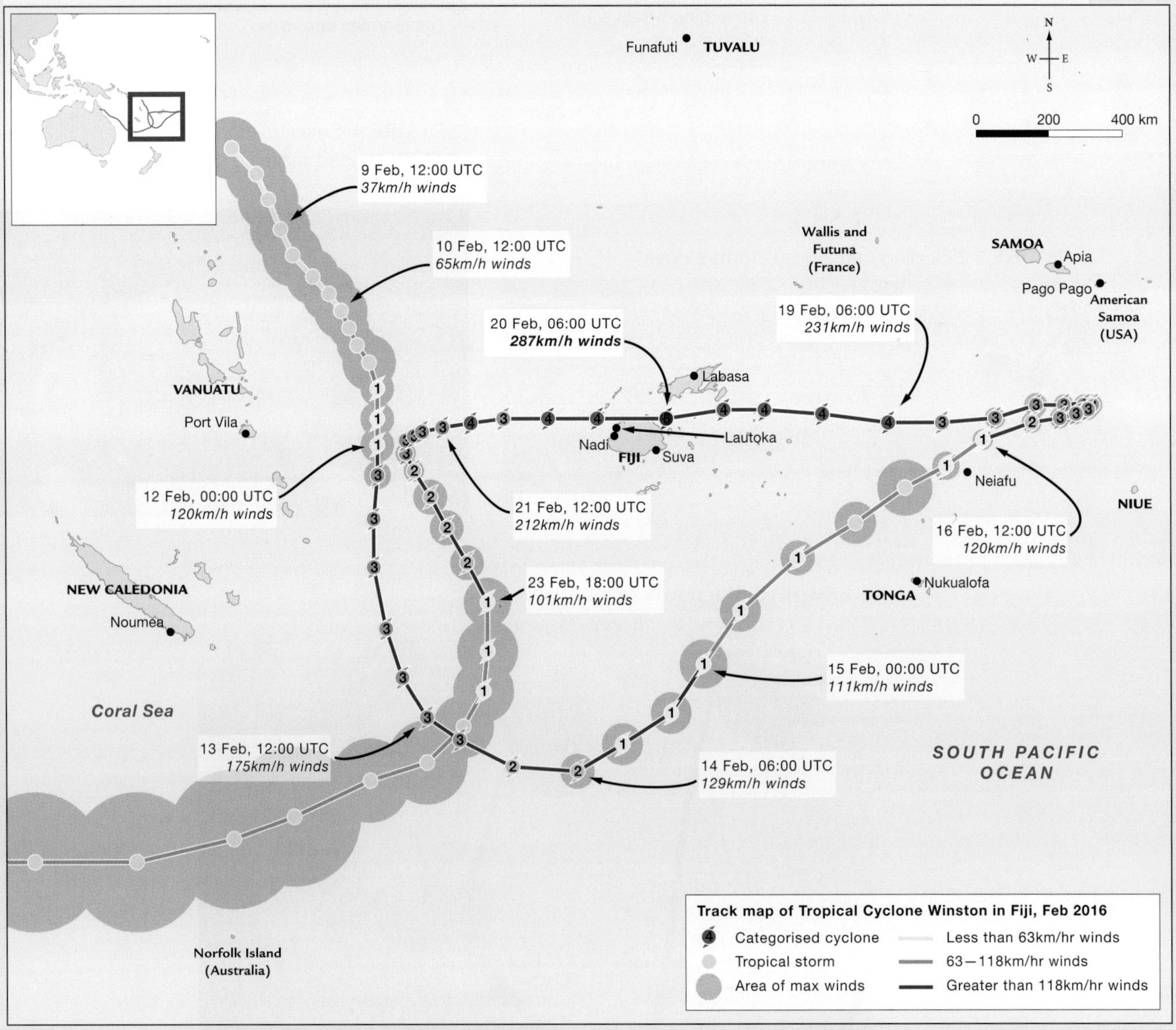

Source: National Hurricane Center, National Oceanic and Atmospheric Administration

Homes and community facilities were flattened in many communities, with some villagers losing all their possessions (see figure 8). Large regions were left without electricity and water. A week after the cyclone around 45 000 people were still living in evacuation centres.

Fiji's largest industries are sugar cane and tourism. The sugar cane industry alone suffered around US$83 million worth of loss. This figure does not take into account the more than 200 000 people who depend on this industry for their livelihood. Additionally, thousands of acres of root crops were lost.

The damage to the tourism industry was mixed. While Denaru Island resorts were still able to operate, this was not the case on some of the outer islands. Despite their losses, many of the local villages that depend on tourism were encouraging tourists to return and they were still operating.

FIGURE 7 Additional damage was caused by a 4-metre storm surge.

FIGURE 8 Whole communities were left devastated.

What aid has Australia provided?

Both Australia and New Zealand were quick to provide assistance to Fiji. Australia worked not only with the Fijian government, but also with the island nation of Tonga, which was also affected by Cyclone Winston (see figure 9).

Were other areas affected?

The east coast of Australia experienced large waves in the wake of Cyclone Winston, forcing the closure of some popular tourist beaches. Despite warnings from authorities, surfers risked serious injury and even death to take advantage of the huge swells created along the New South Wales and Queensland coastlines. Beaches were still closed a week after Fiji was devastated.

FIGURE 9 Australia's aid operation

Australian Government
Department of Foreign Affairs and Trade

Australian Aid

The Australian Government has provided

$15 million in response to #TCWinston in Fiji

Australian support will reach up to **200,000 people.**
This includes humanitarian emergency supplies for over **100,000 people** provided through **Red Cross, UN agencies and NGOs** and delivered by the ADF

Our assistance will help to **restore education and health services** and support livelihoods for those affected by the cyclone.

Australian Medical Assistance Teams are providing lifesaving healthcare in affected communities

ADF is providing:

HMAS Canberra
deployed with 60 tonnes of emergency supplies and personnel to repair critical infrastructure

7 MRH-90 Helicopters
delivering personnel and essential supplies to remote localities

Airlifts
from Australia delivering humanitarian support

Follow @AusHumanitarian on Twitter for updates on Australia's humanitarian response in Fiji

UPDATED 9 Mar 2016

5.6 Activities

To answer questions online and to receive **immediate feedback** and **sample responses** for every question, go to your learnON title at www.jacplus.com.au. *Note:* Question numbers may vary slightly.

Remember

1. What conditions do tropical cyclones need in order to develop?
2. What names are given to tropical cyclones in other *places*?
3. Create a timeline for Cyclone Winston from tropical disturbance to dissipation.

Explain

4. Why do tropical cyclones die out if they move inland?
5. Explain the *changes* that a storm surge can cause to a coastal area.
6. How does the *scale* of a cyclone vary?
7. What is the *interconnection* between the warmth of seawater and cyclones?
8. Study the track map of Cyclone Winston (figure 6). Use this map to describe the *scale* of the cyclone and the damage that was caused.
9. Explain why Cyclone Winston had an impact on more than one *place*.

Predict

10. Describe how the damage would differ between a category 1 and category 5 cyclone.
11. Suggest why people are more likely to be killed or injured after the eye of the cyclone has passed.
12. Cyclones are associated with destructive winds and the displacement of large volumes of water. Which of these events do you think would cause the most damage to the natural and built *environment*? Justify your answer.

Think

13. Refer to figure 1, showing the world pattern of tropical cyclones over *space*.
    (a) When do most cyclones occur north of the equator? When do most cyclones occur south of the equator? Suggest a reason for this difference.
    (b) Name the parts of Australia most at risk from cyclone activity.
14. If the water source for cyclones is the ocean over which they form, explain why strong winds and flooding occur in *places* inland from the coast.
15. Why could we consider that tropical cyclones are an example of the water cycle at work? Give reasons for your answer.
16. Why do you think Cyclone Winston was able to develop into a much stronger storm rather than dissipate once it had impacted on Tonga?
17. Explain the *interconnection* between Cyclone Winston and large waves that resulted in Australian beaches being closed in Queensland and New South Wales.

 RESOURCES — ONLINE ONLY

✦ **Try out this interactivity:** Spiralling sea storm (int-3087)

5.7 What impact did Cyclone Yasi have?

Access this subtopic at **www.jacplus.com.au**

5.8 SkillBuilder: Creating a simple column or bar graph

WHAT ARE COLUMN OR BAR GRAPHS?

Column graphs show information or data in columns. In a bar graph the bars are drawn horizontally and in column graphs they are drawn vertically. They can be hand drawn or constructed using computer spreadsheets.

Go online to access:

- a clear step-by-step explanation to help you master the skill
- a model of what you are aiming for
- a checklist of key aspects of the skill
- a series of questions to help you apply the skill and to check your understanding.

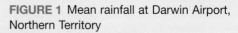

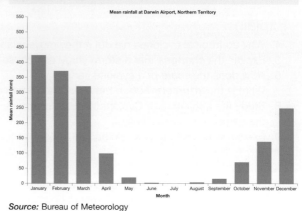

FIGURE 1 Mean rainfall at Darwin Airport, Northern Territory

Source: Bureau of Meteorology

5.9 What are the impacts of typhoons in Asia?

In 2015, 18 typhoons were recorded in the Asian region. Of these, eight reached category 4 or 5 status. While some of these remained at sea, the collective damage bill was US$10.2 million and they resulted in 254 deaths. The strongest was typhoon Soudelor with sustained wind speeds of 215 kilometres per hour and gusting to 285 kilometres per hour. Throughout the typhoon season, two or three active systems were recorded each month, although not all made landfall. At season's end, 254 fatalities had been recorded and the combined damage bill was estimated at US$10.2 billion.

5.9.1 What happened in August 2015?

In August 2015, three typhoons impacted the northern Pacific region. Typhoon Soudelor carved a path of destruction through the region from the 31 July through to 12 August when it finally died out. At the same time, twin typhoons which would subsequently be named Goni and Atsani were already forming. Typhoon Goni would again ravage places still recovering from Soudelor. Typhoon Atsani, however, would remain at sea (see table 1).

TABLE 1 In August 2015 three typhoons formed in the same region of the northern Pacific region.

Typhoon	Active	Area impacted	Strongest winds (kp/h)	Fatalities	Damage ($US)
Soudelor	31 July to 12 August 2015	Mariana Islands, Taiwan, Eastern China, Japan, South Korea, Philippines	285	38	3.72 billion
Goni	13 to 30 August 2015	Mariana Islands, Taiwan, Japan, Philippines, Russia, China, Korea	215	34	293.3 million
Atsani	14 to 25 August 2015	Northern Pacific Ocean	260	n/a	n/a

Typhoon Soudelor

Originally detected as a **tropical disturbance** north of the Marshall Islands on 28 July 2015, within two days after the development of intense swirling thunderstorm activity the system was upgraded to a **tropical depression**. Significant sea surface temperatures of 32 °C saw the system intensify rapidly and officially upgraded to typhoon status on 31 July. Over the next two weeks the system weakened several times as it crossed land, only to re-form and intensify as it moved back over water. Its impact was rated as severe in the northern Mariana Islands, Taiwan and eastern China with 38 confirmed deaths. Lesser impacts were recorded in Japan, South Korea and the Philippines.

FIGURE 1 Starting the clean-up after Typhoon Soudelor

When Typhoon Soudelor made landfall at Saipan in the Mariana Islands it was considered a relatively small system due to the small diameter of its eye at just eight kilometres. However, despite its apparent small size it is evident that it was fed by much stronger winds than at first thought. Accurate wind speed readings are not available, however, as wind-recording instruments in the area broke when the wind speed surpassed 146 kilometres per hour.

The national weather service on Guam estimates that wind speeds were in excess of 200 kilometres per hour, making it a category 3 or 4 typhoon. In the absence of accurate wind readings scientists have used data from weather satellites, Doppler radar, photographs and visual observation of the damage that was caused.

Throughout the region significant damage was caused to infrastructure with winds bringing down power lines and trees, flipping cars and rendering roads impassable. Landslides, mudslides and flash flooding added to the problem. Many places recorded their highest rainfall totals in over 100 years. Refugee centres were filled to capacity as thousands sought refuge from the storm. Around half a million homes were either destroyed outright or suffered significant structural damage.

Downed power lines and snapped wind turbines resulted in millions of households being left without power for weeks. In Taiwan, almost half a million households were left without water when the Nanshi River overflowed and contaminated water supplies, leaving the regional water purification system unable to cope. Primary industries were almost wiped out as significant damage was caused to agriculture throughout the region.

Typhoon Goni

Typhoon Goni struck the Mariana Islands two weeks after they had been devastated by Typhoon Soudelor. As a result, impact was minimal as the clean-up and reconstruction of homes and infrastructure had barely begun. However Goni had a very different track path, with the Philippines, Japan and Korea — which had only been mildly impacted by the previous typhoon — suffering considerable damage.

Torrential rain, flash flooding, mudslides and landslides caused considerable damage to infrastructure and resulted in mass evacuations. Downed power lines left millions without power, and sanitation and water supply were also severely impacted.

Reports out of North Korea indicated that more than 1000 homes and 100 public buildings were destroyed, with 120 hectares of agricultural land flooded. While in Japan more than 600 000 residents had to be evacuated and 70 people were injured. At the same time, Japan was also on high alert in anticipation of a volcanic eruption. Meanwhile, in Russia, which did not bear the full brunt of the typhoon, there was reported damage to 600 buildings and the loss of 88 000 hectares of crops.

FIGURE 2 Using farm machinery to navigate the floodwaters after Typhoon Goni

Typhoon Atsani

Typhoon Atsani developed at about the same time as Typhoon Goni and initially developed into a more powerful system, attaining category 5 status. However, after developing quickly the system also lost power quickly due to significantly dry air above despite the warm ocean temperatures below. After initially following a similar track path as Typhoon Goni, Typhoon Atsani changed direction and headed in a north-easterly direction, coming no closer than 1600 kilometres off the Japanese coast. As a result, only shipping channels in the north Pacific Ocean were impacted.

FIGURE 3 When this image was captured, Goni was a category 4 typhoon and Atsani a category 3. Goni had already impacted on the Mariana Islands. Both systems would later be upgraded to category 5 storms, with Atsani remaining at sea. Goni weakened to category 4 as it approached the Philippines and was downgraded to a tropical storm before it reached Japan.

Source: NASA Earth Observatory

5.9 Activities

To answer questions online and to receive **immediate feedback** and **sample responses** for every question, go to your learnON title at www.jacplus.com.au. *Note:* Question numbers may vary slightly.

Remember

1. What is a typhoon?
2. Which *places* were affected by typhoons Soudelor, Goni and Atsani? Describe how they were affected.
3. Does a typhoon need to cross the coastline to cause damage or loss of life? Explain.

Explain

4. Describe the *scale* of the three typhoons described. Why did they cause devastation to a wide area spanning several countries?
5. Typhoons weaken once they cross the coast. What major threat would be posed to *environments* away from the coast?
6. (a) In groups of three or four, use the map in figure 1 in subtopic 5.6 to make a list of the countries most at risk from cyclones, hurricanes and typhoons. Thinking about the impact of heavy rainfall, storm surges and flooding, select the country that you think might be most affected in terms of economic, social and *environmental* impacts. Use the internet to test your theory and outline these impacts. Present your findings to the rest of the class.
   (b) Evaluate your performance as a team member and assess how well you supported other members of your team. Write a short reflection on your performance.

Think

7. (a) Suggest at least two reasons why the typhoons discussed in this subtopic had different wind speeds and impacts.
   (b) Explain why typhoons might weaken and re-form several times on their journey.
8. Describe the *interconnection* between Typhoons Soudelor and Goni and the Mariana Islands, Taiwan, China and Japan.

5.10 How do tornadoes impact people and the environment?

5.10.1 What is a tornado like?

Tornadoes (or twisters) are violent, wildly spinning columns of air that drop down from under a cumulonimbus cloud and make contact with the ground. They are different from cyclones in that they form over land rather than over water. Unlike cyclones, they are not dependent on a supply of warm water to keep them going. Use the **Tornadoes 101** weblink in the Resources tab to watch a video about how tornadoes form, the damage they cause and how to survive them.

Tornadoes are ranked using a scale called the Enhanced Fujita scale (see table 1). Five categories of wind speed are estimated, based on the damage left behind. These

FIGURE 1 The anatomy of a tornado.

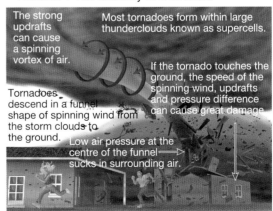

The strong updrafts can cause a spinning vortex of air.

Most tornadoes form within large thunderclouds known as supercells.

If the tornado touches the ground, the speed of the spinning wind, updrafts and pressure difference can cause great damage.

Tornadoes descend in a funnel shape of spinning wind from the storm clouds to the ground.

Low air pressure at the centre of the funnel sucks in surrounding air.

are not wind speed measurements, because most wind speed measuring devices are destroyed during tornadoes, and because the tornadoes die out so quickly.

Widely used in the United States and Canada, the Enhanced Fujita scale is a new version of the original Fujita scale and was designed to expand the descriptions of damage caused at different wind speeds. It also includes descriptions of damage to both the natural and human environments. Within the human environment, the Enhanced Fujita scale includes reference to differences in the construction quality of buildings.

TABLE 1 The Enhanced Fujita scale, like the Beaufort scale, links tornado categories to the damage caused.

Scale wind speed (km/h) category	Typical damage	3-second wind gust speed (km/h)
E0 64–116 Gale	Some damage to chimneys; branches broken off trees; shallow-rooted trees pushed over; signboards damaged.	72–125
E1 117–180 Moderate	Peels surface off roofs; mobile homes pushed off foundations or overturned; moving autos blown off roads.	126–188
E2 181–252 Considerable	Roofs torn off frame houses; mobile homes demolished; train carriages overturned; large trees snapped or uprooted; light-object missiles generated; cars lifted off ground.	189–259
E3 253–331 Severe	Roofs and some walls torn off well-constructed houses; trains overturned; most trees in forest uprooted; heavy cars lifted off the ground and thrown.	260–336
E4 332–418 Devastating	Well-constructed houses levelled; structures with weak foundations blown away some distance; cars thrown and large missiles generated.	337–420
E5 419–512 Incredible	Strong frame houses levelled off foundations and swept away; automobile-sized missiles fly through the air in excess of 100 metres (109 yards); trees debarked.	421+

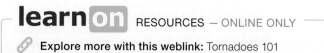

learn on RESOURCES — ONLINE ONLY

Explore more with this weblink: Tornadoes 101

5.10.2 Where is Tornado Alley?

Tornadoes can occur anywhere, but most occur during spring and summer in a part of the United States known as **Tornado Alley** (see figure 2). The worst tornado on record was the Tri-State tornado in March 1925. It destroyed towns across Missouri, Illinois and Indiana, killing 689 people.

The year 2011 ranks third in the history of the United States for the greatest number of strong to violent tornadoes. This number was surpassed only in 1974 and 1965.

FIGURE 2 A map of Tornado Alley, showing the areas of highest risk and high risk

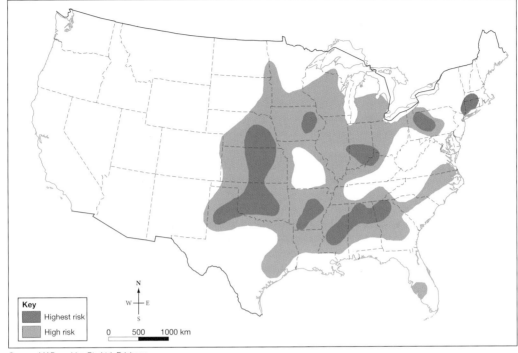

Key
■ Highest risk
■ High risk

0   500   1000 km

Source: MAPgraphics Pty Ltd, Brisbane

Living in Tornado Alley

The people who live in Tornado Alley are well aware of the potential disaster that they face each year during spring and summer. Building codes have been strengthened, requiring all new buildings to have strong roofs and foundations that are tethered to the structure. Most neighbourhoods have early-warning sirens that sound when a tornado is imminent. Most homes have basements or underground **storm shelters** that provide protection for people during a tornado.

FIGURE 3 Joplin, Missouri, where 158 people were killed in May 2011, after a tornado carved a 22-kilometre path of destruction through the town

5.10 Activities

To answer questions online and to receive **immediate feedback** and **sample responses** for every question, go to your learnON title at www.jacplus.com.au. *Note:* Question numbers may vary slightly.

Remember

1. When are tornadoes most likely to occur? Give reasons for your answer.
2. What is Tornado Alley and where is it?

Explain

3. Refer to subtopic 5.6. Explain the difference between a cyclone and a tornado.
4. Refer to figure 1. What is the *interconnection* between large thunderclouds and tornadoes?
5. Do you think we have tornadoes in Australia? Explain.

Discover

6. Tornado Alley is well known, but it is not the only *place* in the world where tornadoes occur. Use the internet to find other locations where tornadoes occur regularly. Investigate one region and one tornado that has taken place. Using ICT, make a short movie clip of the event. Include information about the *scale* of the tornado. You must also cover the aftermath, detailing with the impact on both the natural and built *environments*.

Think

7. Study the graph of tornado frequency in figure 4.
   (a) During which hours of the day are tornadoes more likely?
   (b) Suggest a reason for this pattern.
8. Imagine you are equipping a storm shelter. What 10 items could your shelter not do without? Justify your choices.
9. Refer to figure 2. How does the scale of risk of tornadoes occurring vary within the United States?

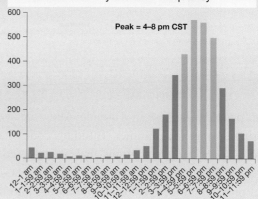

FIGURE 4 Hourly tornado frequency

Peak = 4–8 pm CST

5.11 What was the impact of the May 2013 tornado in the United States?

Access this subtopic at **www.jacplus.com.au**

5.12 What happens when water turns to ice and snow?

5.12.1 What is a blizzard?

Periods of intense snowfall characterised by high winds and snow are known as snowstorms and can be just as deadly as any other storm. The most dangerous snowstorm of all is the blizzard.

The difference between a snowstorm and a blizzard is the strength of the wind. A snowstorm is officially recognised as a blizzard when wind speed is sustained above 56 kilometres per hour or has frequent gusts in excess of this speed for more than three hours. Visibility in a blizzard is also reduced to less than 400 metres. In the most extreme cases it may be difficult to see beyond a metre ahead. Often snow does not fall during a blizzard, but is blown into snowdrifts capable of burying people and objects.

What causes blizzards?

Variations in air pressure (see section 5.2.2) cause strong winds when warm air and cold air meet. It is these strong winds and cold conditions that cause a blizzard to develop.

FIGURE 1 How a blizzard forms

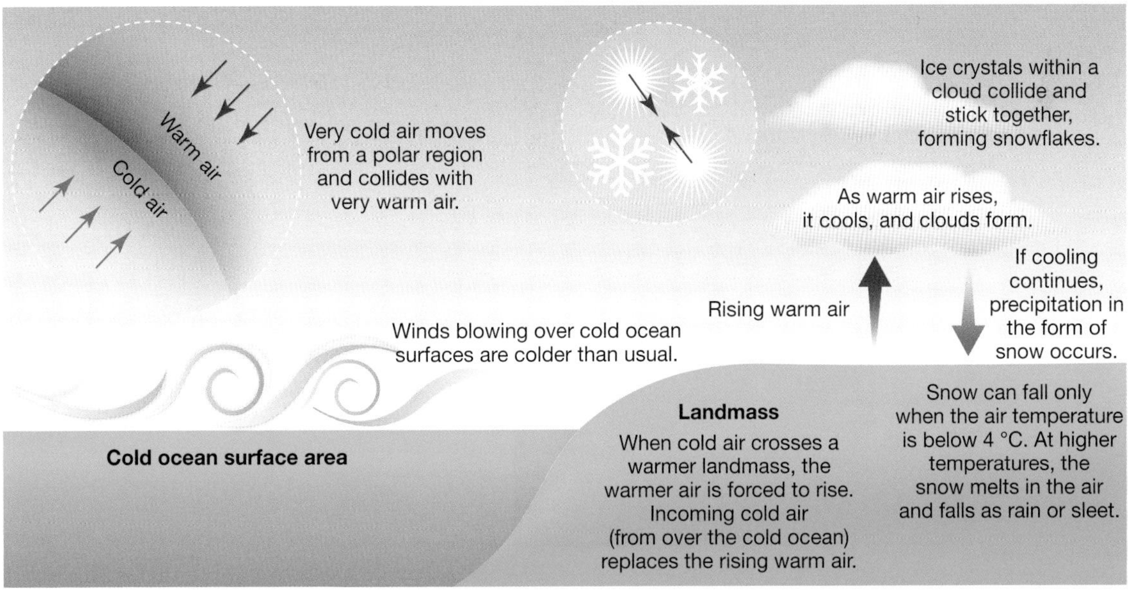

Very cold air moves from a polar region and collides with very warm air.

Ice crystals within a cloud collide and stick together, forming snowflakes.

As warm air rises, it cools, and clouds form.

Rising warm air

If cooling continues, precipitation in the form of snow occurs.

Winds blowing over cold ocean surfaces are colder than usual.

Landmass
When cold air crosses a warmer landmass, the warmer air is forced to rise. Incoming cold air (from over the cold ocean) replaces the rising warm air.

Snow can fall only when the air temperature is below 4 °C. At higher temperatures, the snow melts in the air and falls as rain or sleet.

Cold ocean surface area

Why are blizzards dangerous?

During snowstorms, snow can pile up, and it is impossible to know the depth of the snow, making it difficult to move about. There is the risk of falling through thin ice or into deep **crevasses**. Snow also tends to pile up on slopes. Where the snow load is greater than can be supported by the slope, there is a risk of **avalanches** (see figure 2). An avalanche can be triggered by an earthquake or loud noises such as those produced by a gunshot or by animals.

FIGURE 2 An avalanche

During blizzards a condition known as a **whiteout** can occur (see figure 4). This means there is so much snow that visibility is severely affected and may be limited to just one metre. People and animals cannot tell the difference between the Earth and the sky, and quickly become disoriented, lose their way, and risk freezing to death.

In the extreme cold associated with snowstorms and blizzards, people are at increased risk of **hypothermia**, **frostbite** and suffocation.

What was the world's deadliest blizzard?

The world's deadliest blizzard occurred in Iran in February 1972. A week of low temperatures and strong winds dumped more than three metres of snow and resulted in 4000 deaths. The weather conditions that led to this blizzard are shown in the satellite image in figure 3.

How can buildings be adapted to blizzards?

Researchers in Antarctica have to contend with snow build-up in some parts of the continent. The Halley VI facility (see figure 5) has been built on steel legs that can be raised. Skis have been attached to these legs, so that the entire station can be moved in order to eliminate the dangers associated with accumulating snow.

FIGURE 3 Weather conditions during the 1972 blizzard in Iran

FIGURE 4 Whiteouts reduce visibility, making it easy to become disoriented.

FIGURE 5 The Halley VI Research Station in Antarctica

CASE STUDY

On 22–24 January 2016, a major blizzard dumped 91 centimetres of snow on parts of the mid-Atlantic and north-east United States. The snowstorm covered approximately 1.125 million square kilometres and impacted more than 102 million people. It was officially rated as a category 4 snowstorm for the south-east and category 5 in the north-east (see figure 6). It is among the most powerful storms of all time. Snowfall records were set in a number of cities, including parts of New York, Pennsylvania and North Carolina.

TABLE 1 The Northeast Snowfall Impact Scale combines data on the area covered, the amount of snowfall and the population of the area.

Category	NESIS value	Description
1	1–2.499	Notable
2	2.5–3.99	Significant
3	4–5.99	Major
4	6–9.99	Crippling
5	10.0+	Extreme

FIGURE 6 A polar vortex was responsible, bringing record snow falls to the United States.

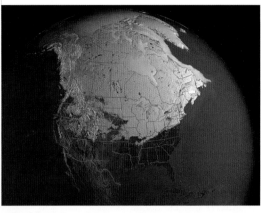

What caused such a severe blizzard in January 2016?

The January blizzard began as an atmospheric disturbance on 20 January that developed into a weak low pressure system before intensifying and triggering a series of thunderstorms. Fed by a **polar vortex** (see figure 6), air pressure continued to drop and the system developed into a major storm bringing freezing rain,

FIGURE 7 Eleven states in the United States were impacted by the blizzard in January 2016.

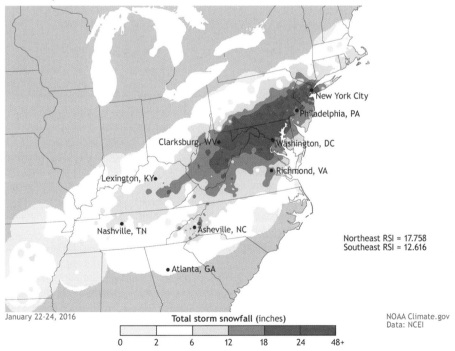

sleet and heavy snowfalls as it moved in a north-easterly direction across the United States (see figure 8). The remnants of the storm were felt as far away as the United Kingdom before it finally dissipated over Finland on 29 January. At its peak, the storm system recorded a central air pressure of 983 millibars.

How did this blizzard impact people and places?

A state of emergency was declared in Washington DC and 11 other states as authorities prepared for record-breaking snowfalls. More than 100 million people were affected with 33 million placed on blizzard watch. Around 500 000 people suffered power outages. The National Guard was placed on standby and crews were brought in from other parts of the country to assist power companies check lines and restore power. Around 1.5 million tonnes of road salt was distributed across the region and more than 7000 pieces of snow equipment were mobilised to help deal with the aftermath. People were urged to stay indoors and off the roads.

Records tumbled across the United States. The highest falls were recorded at Mount Mitchell in North Carolina with 170 centimetres of snow, setting a new all-time record (see table 2). Daily snowfall records were also recorded in several places.

The blizzard caused major disruptions for people travelling by air, with the travel plans of more than 100 000 travellers thrown into chaos (see figure 8). More than 10 000 flights were cancelled in the United States, many of these due to the 'ripple effect' caused by the cancellation of flights into and out of airports in the affected area. The number of cancelled flights also 'rippled' internationally, with around 200 flights in Canada, Mexico and the United Kingdom also cancelled.

It is estimated that the damage bill could be as high as US$3 billion. Fifty-five fatalities were also recorded; some of which have been directly attributed to heart attacks caused by shovelling snow. In Washington DC, police issued almost $1.1 million worth of parking fines and $65 000 in fines for cars abandoned on snow emergency routes. More than 700 vehicles were towed by authorities and impounded.

TABLE 2 All-time record snowfalls

Location	State	Amount
Mount Mitchell	North Carolina	170 cm
Allentown	Pennsylvania	81 cm
Philadelphia International Airport	Pennsylvania	77 cm
John F Kennedy International Airport	New York	77 cm
Baltimore-Washington International Airport	Maryland	74 cm
Newark	New Jersey	71 cm

FIGURE 8 Flights across the affected region were grounded after record snow falls.

Why is shovelling snow dangerous?

In the United States, more deaths occur during blizzards than from hypothermia and motor vehicle accidents combined.

Every winter around 100 people in the United States die from heart attacks caused by shovelling snow. Researchers have found that shovelling snow places more strain on the heart than a vigorous session on a treadmill (see figure 9). Both heart rate and blood pressure increase more dramatically when using arm muscles as opposed to leg muscles. Most people shovel snow in the early morning when the temperature is at its coldest, causing arteries to constrict which in turn decreases blood supply which in turn leads to cardiac arrest.

FIGURE 9 Shovelling snow increases the risk of heart attack.

FIGURE 9 Shovelling snow increases the risk of heart attack.

5.12 Activities

To answer questions online and to receive **immediate feedback** and **sample responses** for every question, go to your learnON title at www.jacplus.com.au. *Note:* Question numbers may vary slightly.

Remember

1. How are snowstorms and blizzards different?
2. Why are whiteouts so dangerous?

Explain

3. Describe the visible *changes* to the *environment* that a blizzard brings.
4. Where is the Halley VI Research Station? Explain why it has been built on skis.

Predict

Study figure 10. Who do you think is most at risk from the extreme cold? Give reasons for your answer.

Think

5. Describe the *interconnections* between the atmosphere and the land that cause blizzards.
   (a) Refer to the work on natural hazards you have completed in other sections. Explain which hazard you believe would be the most dangerous. Explain why you think this.
   (b) Survey your class to find out which hazard is voted the most dangerous.
   (c) As a class, compile a list of reasons why class members voted for their chosen hazard.
6. With the aid of a diagram, explain the operation of the water cycle in colder regions.

FIGURE 10

5.13 Are there water hazards in Mongolia?

Access this subtopic at **www.jacplus.com.au**.

5.14 How do we respond to extreme weather and water events?

5.14.1 How do I prepare?

With today's modern technology we have access to a wealth of information that enables individuals and communities to prepare themselves for the wild winds over which they have no control. In many cases, the winds also bring vast amounts of rainfall and the land is often **inundated**. While we can in some ways prepare for such events, it is inevitable that both the natural and built environment will be impacted.

People who live in disaster-prone areas should know the risks associated with the potential hazards they face and the time of the year when they are at greatest risk. In Queensland, for example, where tropical cyclones bring flooding rains, houses are often built on stilts.

The key to survival is to be prepared. Securing your home and having an emergency kit are two important things that can be done on a continual basis.

5.14.2 What happens in Chatham County?

Chatham County is located in the US state of Georgia. It has a population of approximately 270 000 and occupies an area of 1637.6 square kilometres.

On average, Chatham County expects up to 19 severe thunderstorms each year. As Georgia lies in Tornado Alley, people in Chatham can also expect an average of six tornadoes each season. In order to assist residents, a network of early warning sirens has been installed in places where people typically tend to gather. Ninety-five per cent of the county is covered. When the sirens give out their distinctive, extended wail, residents know to seek shelter. Most homes and public buildings in the county are equipped with storm cellars.

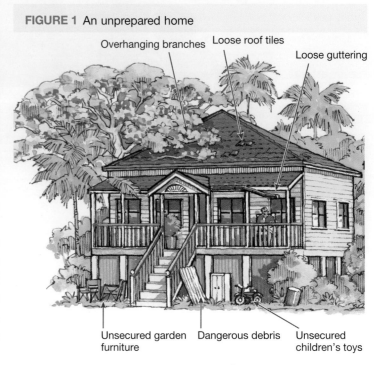

FIGURE 1 An unprepared home

Overhanging branches

Loose roof tiles

Loose guttering

Unsecured garden furniture

Dangerous debris

Unsecured children's toys

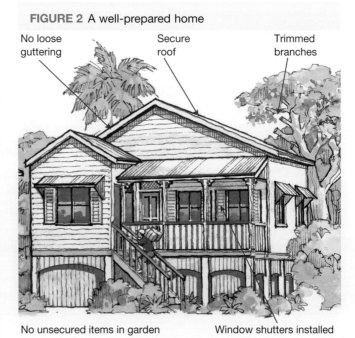

FIGURE 2 A well-prepared home

No loose guttering

Secure roof

Trimmed branches

No unsecured items in garden

Window shutters installed

Did you know?

- Dwellings in regions that are prone to snow and heavy rain have steeply sloping roofs to prevent the build-up of water, ice and snow on the roof. In contrast, houses in hot, dry climates often have flat roofs.
- When cyclones and tornadoes are approaching, you should turn off all power.
- A small room such as a bathroom is the safest place to take refuge if you do not have access to a storm cellar.
- Special building codes exist in areas prone to tornadoes and cyclones, with their associated strong winds.
- Shutters on windows help keep water out and reduce the likelihood of injury from broken glass.

FIGURE 3 An emergency kit

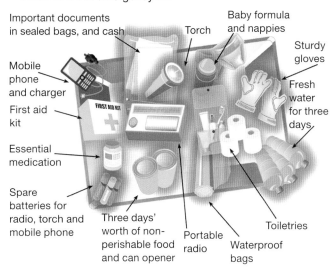

FIGURE 4 The location of sirens in Chatham County

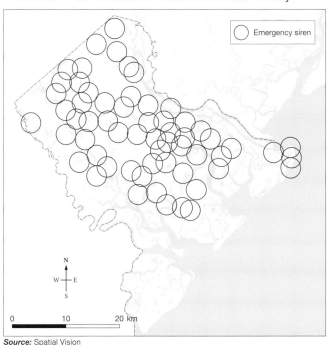

Source: Spatial Vision

FIGURE 5 The entrance to a storm cellar

5.14 Activities

To answer questions online and to receive **immediate feedback** and **sample responses** for every question, go to your learnON title at www.jacplus.com.au. *Note:* Question numbers may vary slightly.

Explain

1. Study figures 1 and 2. Explain to the people who live in the house shown in figure 1 what they need to do to prepare their home for the next cyclone season. Identify particular hazards and the potential risks they pose.
2. Explain why the homes shown in figures 1 and 2 are elevated.

3. What do you think poses the biggest threat — wind or water? Justify your choice.
4. What have the people in Chatham County done to deal with hazards in their *environment*?

Predict
5. What do you think the people who live in figure 2 are preparing for? Give reasons for your answer.

Think
6. The residents of the house shown in figure 2 live in a *place* that has been placed on cyclone alert and have asked you for advice. They have begun preparing an emergency kit like the one shown in figure 3. Other than food, what items would you suggest they include?
7. What items do you think you would need in a blizzard that you would not need in a thunderstorm, tornado or tropical cyclone? Explain.
8. Which regions within Chatham County are not well covered by sirens? Use evidence from the map to suggest a reason for this.

5.15 Review

5.15.1 Review
The Review section contains a range of different questions and activities to help you revise and recall what you have learned, especially prior to a topic test.

5.15.2 Reflect
The Reflect section provides you with an opportunity to apply and extend your learning.
Access this subtopic at **www.jacplus.com.au**

TOPIC 6
Fieldwork inquiry: What is the water quality of my local catchment?

6.1 Overview

Numerous **videos** and **interactivities** are embedded just where you need them, at the point of learning, in your learnON title at www.jacplus.com.au. They will help you to learn the content and concepts covered in this topic.

6.1.1 Scenario and your task

Water is our most valuable resource, and the management of this vital resource should be a priority at a local, regional and global scale. Everybody lives near a catchment — there is usually a river, creek, drain or other waterway close to your home, school or neighborhood. If you live in an urban area, the creek may have been highly modified and may look like a concrete drain. Water quality in these waterways will vary from place to place and is influenced by many factors.

Water quality can affect health in many ways. Rivers and streams act as drainage systems. When it rains, water transports rubbish, chemicals and other waste into drains and, eventually, rivers.

Your task

Your team has been selected to research the water quality of a local catchment or waterway and produce a report and presentation on your findings. Be sure to measure water quality at different locations along the river, creek or stream, and try to determine the causes of different water quality.

6.2 Process

6.2.1 Process

- You can complete this project individually or invite members of your class to form a group.
- **Planning:** You will need to research the characteristics of your local catchment area. In order to complete sufficient research, you will need to visit a number of sites within the catchment, comparing

different locations upstream and downstream of one creek or river. Research topics have been loaded in the Resources tab to provide a framework for your research:

– **What** sort of data and information will you need to study water quality at your fieldwork sites?
– **How** will you collect and record this information?
– **Where** would be the best locations to obtain data? You can determine this once you know which waterway(s) you are visiting.
– **How** will you record the information you are collecting? Consider using GPS, video recorders, cameras and mobile devices (laptop computer, tablet, mobile phone).

6.2.2 Collecting and recording data

It is important that you have some knowledge of the fieldwork location before you visit the site. Access to topographic maps and Google Earth will help you become familiar with the location. Using these tools, complete a sketch map of the waterway(s) and label the sites you are going to visit. You can then scan your sketch map and have it available electronically on the field trip. Alternatively, use Google Maps to record all the sites you visit. Ensure that you bring all the equipment and resources you will need to collect the data with you to each site. It would be useful to work in groups to collect the data, with each group collecting different data at each site. Use the supplied data collection templates electronically on your mobile device, or print copies.

6.2.3 Analysing your information and data

Once you have collected, collated and shared your data, you will need to decide what information to include in your report and the most appropriate way to show your findings. If using spreadsheet data, make total and percentage calculations. Some measurements are best presented in a table, others in graphs or on maps. If you have used a spreadsheet, you may like to produce your graphs electronically. Use photographs as map annotations (either scanned and attached to your electronic map or attached to your hand-drawn map) to show features recorded at each site. You may also like

to annotate each photograph to show the geographical features you observed. Describing and interpreting your data is important. There are broad descriptions that can also be made of your findings, which might include:

• Where is water quality highest (best) in the waterway studied?
• Is water quality better in the upper reaches of the river or creek?
• Does an urban waterway have better water quality than a rural waterway?
• Does surrounding land use have an impact on water quality?
• Do large waterways have better water quality than smaller waterways?
• What were the main contributors to poor and good water quality?
• How does surrounding vegetation impact on water quality?

Download the report template and the presentation planning template from the Resources tab to help you complete this project. Use images, videos and audio files to help bring your presentation to life. Use the report template to create your report. Use the presentation template to create an engaging presentation that showcases all of your important findings.

6.2.4 Communicating your findings

You will now produce a fieldwork report and presentation to present your findings. Your report and presentation should include all of the research that you completed and all evidence to support your findings. Ensure that your report includes a title, an aim, a hypothesis (what you think you will find, which is written before you go into the field), your findings and a conclusion. You will also need to recommend some type of action that needs to be taken to improve water quality in the creek or river you visited.

6.3 Review

6.3.1 Reflecting on your work

Think back over how well you worked with your group on the various tasks for this inquiry. Determine strengths and weaknesses and recommend changes if you were to repeat the exercise. Identify one area where you were pleased with your performance, and an area where you would like to improve. Write two sentences outlining how you might be able to do this.

Print out your Research Report and hand it in with your fieldwork report and presentation, and reflection notes.

UNIT 2
PLACE AND LIVEABILITY

Have you ever stopped to consider why you live where you do? What prompted your family to live there? There are so many different types of places where you *could* live: rural or urban, coastal or inland, small or large, bustling or quiet. Different people find different places suitable (or more 'liveable') for them than other places. Some people have no choice. The question is: how can we make places more liveable?

Many people like to travel to beaches like this popular one in Scheveningen, Holland, or to overseas cities, to experience differences other places offer.

TOPIC 7
Where do Australians live?

7.1 Overview

Numerous **videos** and **interactivities** are embedded just where you need them, at the point of learning, in your learnON title at www.jacplus.com.au. They will help you to learn the content and concepts covered in this topic.

7.1.1 Introduction

Why does your family live in the place, state, city, street or house that it does? Why do many Australians live in big cities near the coast? Have you thought about the reasons why your parents selected the place or environment in which you now live? People living in Australia have been making choices about where to live for many thousands of years. Has the region of Australia around Port Jackson in modern Sydney always been the most heavily populated part of Australia? Do people choose to live in places they feel are the most liveable? Let's try to work out why Australians choose to live in the places they do.

Starter questions

1. If you could live anywhere in the world, where would it be and why?
2. If your answer to question 1 is not in Australia, which part of Australia do you think is the most similar to the *place* you chose?
3. What are the geographic features of the place you would ideally like to live in? Geographic features can usually be mapped, and include climate, landscape, *environment* (either built or natural), jobs, culture, infrastructure, wealth and safety. Find an image of this *place* and annotate its geographic features.
4. Look at the image on this page. List the positive and negative aspects of living in this *place*.

Would you like to live here?

7.2 What creates a sense of place?

7.2.1 A sense of place

Places are central to the study of geography. This is because geographers are interested in where things are found on Earth and why they are there. But what exactly is a place?

To understand what a place is, think about **location** and **region**. Each place has a unique identity that makes it different from other places. A combination of characteristics is specific to that place, making it individual. A sense of place comes from being aware of what makes that location significant and seeing its special qualities.

The characteristics of a place can come from:
1. natural features
2. human features — that is, built by people
3. a combination of the two.

Eventually, one or more of these features becomes a symbol of that place in people's minds.

FIGURE 1 The Taj Mahal in Agra, India

FIGURE 2 The Grand Canyon, Utah, United States

FIGURE 3 Rio de Janeiro, taken from a helicopter, showing the Corcovado in the foreground with the statue of Christ on it and Sugarloaf Mountain, or Pao de Acucar, in the background, to the right

FIGURE 4 Disney World, Orlando, Florida, United States

FIGURE 5 Table Mountain, Cape Town, South Africa

FIGURE 6 The Golden Gate Bridge, San Francisco Bay, United States

FIGURE 7 The Great Wall of China

7.2 Activities

To answer questions online and to receive **immediate feedback** and **sample responses** for every question, go to your learnON title at www.jacplus.com.au. *Note:* Question numbers may vary slightly.

Remember

1. Study figures 1, 2, 3, 4, 5, 6 and 7. Describe five characteristics in the *environment* of each feature that create its individual sense of *place*. Consider natural as well as human features.
2. Of all these characteristics, which one do you believe to be the most important in creating an identity for that *place* in the minds of people?
3. Suggest reasons why these *places* have become famous around the world.

Discover

4. (a) Conduct a survey of your class to find out each person's top five favourite *places* in Australia. Collate the results in a table like the one below.

Place	Student A	Student B	Student C
Great Barrier Reef	✔		✔
Uluru		✔	✔
My grandparents' farm near Ballina, NSW		✔	

This table could also be set up electronically, using a spreadsheet program.
   (b) Graph the results to show the ranking of the *places* by percentage of the class; for example, 45 per cent of the class named Uluru in their top five *places* in Australia.
   (c) As a class, discuss the patterns shown by the graph. Suggest reasons to explain why people like or dislike certain *places*.

Think

5. Do you think that people's top five favourite *places* would vary with the age of the individual? Explain your answer. (You might need to interview a few people of different ages to help you work out an answer to this.)
6. No matter where we live, we all live in the one *place*: Planet Earth. From what you have learned so far, define what a *place* is in your own words. What do you think would be the characteristics of a *place* that would appeal to anyone, wherever they come from? (*Hint:* What feelings do you have when you are in a *place* that you like?)

 Deepen your understanding of this topic with related case studies and questions. ⊙ **New York**

7.3 Why do people live in certain places?

7.3.1 Push and pull

People choose to live in specific places for a wide range of reasons. These reasons can be broadly divided into **pull factors** and **push factors**. The combination of reasons varies from person to person, and what is an advantage for one person may be seen as a negative by someone else.

It is also true, though, that the reasons people choose to live in a place often change over time. Sometimes, these reasons might even be connected to the very existence of the place — or its changing nature.

There are four factors that influence the liveability of places or why people decide to live there:

1. available resources (money)
2. employment

3. relationships with other people (for example, wanting to be near family or moving for a partner's job)
4. lifestyle.

Many of these factors change throughout a person's life. For example, where a 20-year-old single person wants to live is often quite different from where someone in their forties, or someone with a partner and two teenage children, may want to live.

In other situations, the reason for living in a place may disappear. The town of Rawson, near Mt Erica in Victoria, was built for the people building the Thomson Dam in the 1970s and early 1980s (see figure 1). After the project was finished in 1983, nearly every family left the town because there was no longer any work there. Its **community** identity had to change. The few people left in Rawson now provide services for people using the area for recreations such as bushwalking and skiing.

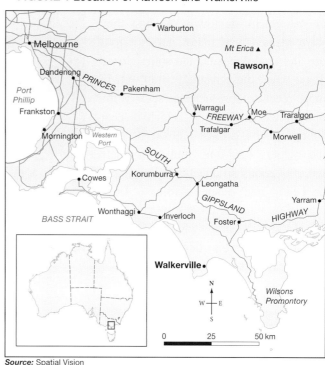

FIGURE 1 Location of Rawson and Walkerville

Source: Spatial Vision

Walkerville is a small coastal settlement on the coast of Victoria, just east of Inverloch and Venus Bay near Wilsons Promontory (see figure 1). Walkerville is a good example of the way people's reasons for living in a place can change over time.

Walkerville was built to provide a place for the workers who were to produce quicklime from the limestone cliffs. Cement was in great demand for building in Melbourne at this time, and lime could be transported there easily by ship. The town itself disappeared when the limestone cliffs were all mined out.

The modern settlement of Walkerville is now a small, isolated holiday location, popular with fishermen, and located next to the Cape Liptrap Coastal Park. Much of the original settlement of Walkerville no longer exists, but the ruins that remain, along with the old cemetery, give us a good picture of what the place used to be like (see figure 2).

FIGURE 2 (a) The Walkerville lime-burning kilns in 1972 (b) The kilns are now a tourist attraction. (c) There are no shops for 20 kilometres in either direction, except for the caravan park kiosk on the foreshore.

(a)

(b)

(c)

Many of the towns in the north-eastern United States were established as manufacturing towns. At first they were located near major ports or iron ore and coal deposits, and some closed down when these resources ran out. In more recent times, factories such as the one shown in figure 3, which is near Baltimore, have closed down because the owners could no longer compete with the goods produced at a lower cost in China and other south-east Asian countries. With no other jobs available, people left the area, which has fallen into a state of **urban decay**.

FIGURE 3 A disused factory near Baltimore

7.3 ACTIVITIES

To answer questions online and to receive **immediate feedback** and **sample responses** for every question, go to your learnON title at www.jacplus.com.au. *Note:* Question numbers may vary slightly.

Think

1. Study figure 2 (c) and research the *changes* over *time* that have occurred in Walkerville as a *place*. Was the decline of the original township of Walkerville due to push or pull factors? How did these influence people's choice of where they would live? Justify your answer.
2. Identify and justify the push and pull factors that exist for people thinking about whether they should move to Walkerville today.
3. In groups of three, discuss the difficulties that would have been faced by the lime-burners who lived in the original settlement of Walkerville, given its *location.*

Explain

4. Study figure 3. Identify some of the specific signs that indicate an area is in urban decay.
5. Suggest reasons why some people continue to live in decaying urban *environments*, and why others might choose to move.

Discover

6. (a) Survey the members of your class and find out the reasons why their families chose to live in the *place* or *location* where they do. Classify the responses using the four categories named on this spread in a table like the one below.

Student	Resources	Employment	Relationships	Lifestyle
Gina	Near major shops	Near my dad's work		
Miguel		Near my mum's work	Close to my family who came to Australia earlier Close to my father's best friend	
Daniel				Near the sea, as we all sail or surf

(b) Present the answers using a column graph, correctly and fully labelled.
(c) As a class, discuss the pattern of reasons shown by the graph, and the possible explanations for this. For example, how important to people are social connections?

Predict

7. Look up on Google Earth the *location* of the current settlement of Walkerville. Calculate the distance between Walkerville and the settlements around it. Study the land use and features of the environment around the settlement. Identify and list the advantages and disadvantages of Walkerville as a holiday *location,* using evidence from your Google Earth study.

8. A developer has proposed to the local shire council and the state government that the farmland around Walkerville should be rezoned to allow the building of a large holiday resort. In your opinion, would this be a good or bad policy for the future of the residents of Walkerville? Give reasons for your answer, referring to features that you can identify on the Google Earth map.

learn on RESOURCES — ONLINE ONLY

✦ **Try out this interactivity:** Push/pull factors (int-3089)

7.4 Where do you live and why?

7.4.1 How did I get here?

When we first learned to write our address, we often included our house number, street, town, city, state, country, continent, hemisphere, planet and universe. You could also identify your location with GPS coordinates, a grid reference or by use of latitude and longitude. Knowing the place where you are is important but so is how you got there.

FIGURE 1 Australians born overseas

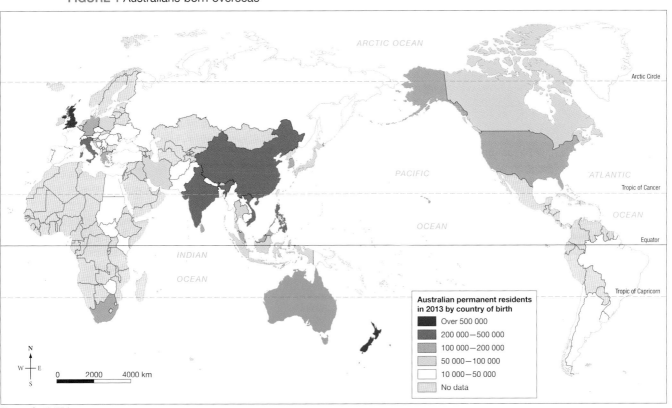

Source: Spatial Vision

A *Cindy:* When I lived in Bejing I was called Jing-Wei. I came to Sydney in 1988 to study economics at university. I became an Australian citizen in 1993. I now have three Australian children.

B *Andrew*: I came to Perth with my wife and three children in 1993 just before Nelson Mandela was elected president of South Africa. We were concerned for our safety in Johannesburg and were keen to start a new life in a country with a similar climate and language. Now two of my brothers also live in Australia.

C *Lucy*: My brother moved from Palmerston North in New Zealand to Melbourne for work in 2006 and I followed him the next year. I like living in a larger city. There is more going on and I get paid a lot more. One day I might return to New Zealand.

D *Deepak*: My family moved from Delhi in 1988 when my father was offered a job in a computer company in Adelaide. There were not many Indian kids in my school but I studied hard and went to university. I now have three children and live in Newcastle.

Did you and your family arrive by boat, plane or car, or were they born here? What decisions were made by your parents or grandparents which resulted in your family living in your place, house, state, country or hemisphere? Over 25 per cent of Australia's population was born overseas, and it is estimated that most will move homes between 11 and 12 times during their lifetime.

What is your story?

Activity question 1 allows you to investigate why you live in your place. It is a task of discovery, and will take you some time to complete. Your aim is to discover your family's migration story and why you live where you do. Does your family have a recent migration story or did your family migrate with the First Fleet? Do you have Aboriginal or Torres Strait Islander heritage? Did your grandparent build the house you live in or did your parents or carers build your house? How has the place your family lives in changed over time and space?

7.4 Activities

To answer questions online and to receive **immediate feedback** and **sample responses** for every question, go to your learnON title at www.jacplus.com.au. *Note:* Question numbers may vary slightly.

Discover

1. This task allows you to discover your family's migration story and why you live where you do.
   Step 1
   What can you find out for yourself?
   Your place
   • What is your address? Write out your full address, including your hemisphere, latitude and longitude.
   • Use Google Maps or Whereis to locate and identify your house in your street. Download an aerial view and a street view of your house.
   • Annotate your aerial photo or map to identify who lives in your house, including pets, and which parts of the house they use. You could illustrate the people who live in your house in a cartoon — like the stickers, or decals, of families that people put on their cars.
   • Ensure that your map has a compass, approximate scale and appropriate title.

Step 2

Your neighbourhood

- Using Google Maps or Whereis, download an aerial view of your street or at least the eight closest houses or dwellings.
- Using the family decals, annotate each house to show who lives in it.

Step 3

How long have you lived at this address?

If you have previously lived somewhere else, list and map your past addresses. How many times have you moved? Share with your class the information that you have collected so far.

Step 4

How did you get here?

To investigate the rest of your story, you will need to speak to your parents and possibly your grandparents. As you collect information about where your parents and grandparents were born, create a family tree of *places*. Try to find out why and when your relatives came to Australia. Figure 2 illustrates how this may look.

- Where were your parents/carers born?
- How did they travel from where they were born to the *place* you now live?
- Why did your parents move to where you now live? Would they prefer to live in another place that is more liveable?
- Why did your grandparents and great-grandparents move from their *place* of birth?

Use the **Family tree template** weblink in the Resources tab to create your family tree.

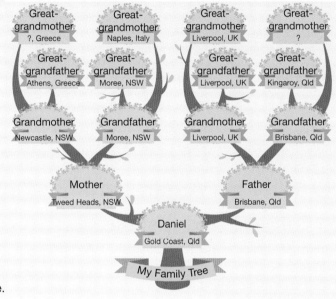

FIGURE 2 An example of what your family tree may look like

Great-grandmother — ?, Greece
Great-grandmother — Naples, Italy
Great-grandmother — Liverpool, UK
Great-grandmother — ?

Great-grandfather — Athens, Greece
Great-grandfather — Moree, NSW
Great-grandfather — Liverpool, UK
Great-grandfather — Kingaroy, Qld

Grandmother — Newcastle, NSW
Grandfather — Moree, NSW
Grandmother — Liverpool, UK
Grandfather — Brisbane, Qld

Mother — Tweed Heads, NSW
Father — Brisbane, Qld

Daniel — Gold Coast, Qld

My Family Tree

FIGURE 3 Immigrants arriving in Australia by plane, 1967

Think

2. Use your family tree, along with the transport your family used, to create a map that shows this *interconnection*.

   Things to think about before starting your map:

   • What *scale* and size of map will you need?
   • Would you be better off having two maps? In the example shown in figure 2, Daniel's parents and grandparents mostly came from New South Wales and Queensland but most of his great-grandparents came from Europe. To map this information, he should use a world map plus a larger *scale* break-out or inset map of New South Wales and Queensland.
   • How will you show the type of transport? Coloured arrows might work well.
   • What is an appropriate title for your map?
   • Would you like to illustrate your map with images of your relatives, their houses, flags of the countries they came from or images of the transport that they used? If you wish to add images, you will need to have a larger *scale* base map than if you just used symbols.
   • You could annotate the map with the reasons your relatives moved.

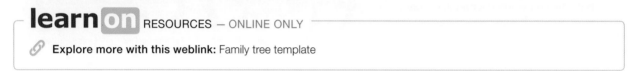

learn on RESORCES — ONLINE ONLY

🔗 **Explore more with this weblink:** Family tree template

7.5 Why do Australians live in remote places?

7.5.1 Settling inland Australia

For over 100 years, a small percentage of Australians have been moving away from large cities and coastal regions to live in more **remote** locations. They are often searching for new farmland or the mineral resources of the inland. Why do some people choose to live in places where their nearest neighbour is 50 kilometres away and it takes six hours to get to the closest supermarket? Why do they find remote places more liveable?

The potential to relocate people inland has never been faster or easier. The interconnection provided by modern transport and the high speed communication provided by phone and internet should mean that technology has reduced remoteness.

The general shift of Australia's population for the last 100 years has been towards the major cities and away from the country. The average age of farmers in Australia is about 53 years and getting older. Most children of farmers leave the country and seek education and work opportunities in large cities. Figure 1 shows how quickly the inland of Australia was occupied after 1825.

Over the past 100 years, there have been many attempts by governments and private industry to encourage people to occupy the more remote places of Australia. Soldier settlement programs and mining developments are two such schemes.

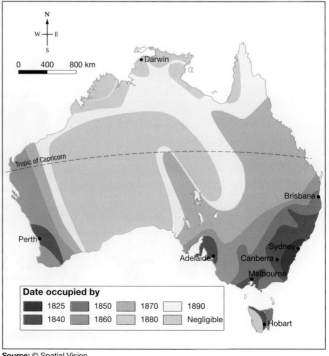

FIGURE 1 Stages in European land occupation in Australia

Date occupied by

1825	1850	1870	1890
1840	1860	1880	Negligible

Source: © Spatial Vision

Soldier settlement schemes

After both World War I and World War II, the state and federal governments of Australia began a program of providing land to returned soldiers. This was to give these soldiers work, but it was also seen as a way of attracting people to otherwise sparsely inhabited places.

After World War I, soldier settlements included Merbein and Mortlake in Victoria, Griffith and Dorrigo in New South Wales, Murray Bridge and Kangaroo Island in South Australia and the Atherton Tableland in Queensland. The settlers were expected to stay on their land for at least five years and to improve the quality of the land they were farming. Many of these settlements were not successful because the soldiers were not always suited to farming, the farms were often too small, and farmers did not have enough money to invest in stock or equipment.

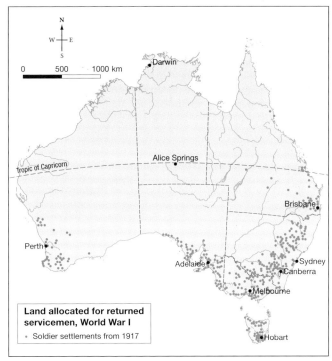

FIGURE 2 Location of soldier settlement areas, 1917

Land allocated for returned servicemen, World War I
· Soldier settlements from 1917

Source: Spatial Vision

After World War II, a similar scheme was much more successful, because farms were bigger, and roads, housing and fences were supplied. Over 25 000 soldiers were resettled after World War I.

Remote mining communities

Karratha Broken Hill and Tom Price are examples of current mining towns that are just as remote as were the goldrush towns of Bathurst and Ballarat in the 1850s and 1860s.

Today it takes less than five hours to fly directly from Brisbane to Tom Price, yet it can be difficult to attract workers to mines in this region. Wages are high; for example, a truck driver can earn $150 000 per year. There are now fewer jobs because the mining boom has passed, but skilled workers are attracted to these remote places. Some workers **fly in and fly out (FIFO)** for their shifts. They live with their families in less remote places such as Brisbane and fly in for a shift that may last several weeks, eventually flying home for their days off.

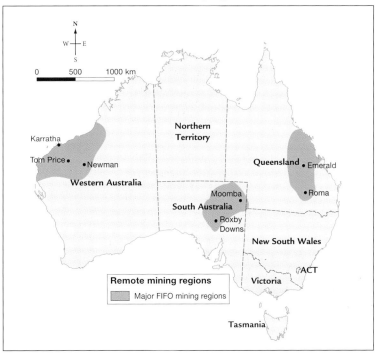

FIGURE 3 Location of remote mining regions

Remote mining regions
▓ Major FIFO mining regions

Source: Spatial Vision

FIGURE 4 Mount Tom Price mine and Tom Price township: note the FIFO workers' huts in the left foreground

7.5 Activities

To answer questions online and to receive **immediate feedback** and **sample responses** for every question, go to your learnON title at www.jacplus.com.au. *Note:* Question numbers may vary slightly.

Remember

1. What makes a *place* remote?
2. How does FIFO reduce remoteness?

Explain

3. Describe the *change* in the speed of settlement of inland Australia that is illustrated by figure 1.
4. (a) Compare figure 2 with figure 1. When were the soldier settlement *places* first settled?
   (b) Use your atlas to compare the location of soldier settlements with a rainfall map of Australia. Were soldier settlements located in *places* that receive good rainfall for farming?
5. The soldier settlements of 1917 were established on remote, underused land. One hundred years later, would these places still be considered remote? Refer to figures 2 and 3 in your answer.

Discover

6. Research a local soldier settlement scheme. When was it established? How successful was it? How did this scheme help to populate a remote *place*? Map its geographic features by using Google Maps.
   Use the **Soldier settlement** weblinks in the Resources tab to help with your research.

Predict

7. How might people be encouraged to move from the coastal fringe to the more remote *places* of Australia? What could make you or your family move or relocate? Produce a short film, snappy slide show, or an advertising campaign that highlights the pull factors which might make people *change* the *place* where they live.

Deepen your understanding of this topic with related case studies and questions.
❯ **Population of Australia**

 RESOURCES — ONLINE ONLY

 Try out this interactivity: Remote living (int-3090)

 Explore more with this weblink: Soldier settlement

7.6 SkillBuilder: Using topographic maps

WHAT ARE TOPOGRAPHIC MAPS?

Topographic maps are a type of map that provides detailed and accurate information of features that appear on the Earth's surface.

They show features of the natural environment, such as forests and lakes, and features of human environments, such as roads and settlements. Relief is often shown using contour lines.

Go online to access:

- a clear step-by-step explanation to help you master the skill
- a model of what you are aiming for
- a checklist of key aspects of the skill
- a series of questions to help you apply the skill and to check your understanding.

FIGURE 1 Topographic map extract of Mount Gambier

learn on RESOURCES — ONLINE ONLY

Watch this eLesson: Using topographic maps (eles-1641)

Try out this interactivity: Using topographic maps (int-3137)

Complete this digital doc: Topographic map extract_Mt Gambier (doc-17951)

7.7 What draws people to rural areas?

7.7.1 Rural settlement

Some people live in rural areas because they are involved in primary industries. Others provide services.

Griffith is a large town (population 17 000) in the Murrumbidgee Irrigation Area in New South Wales. The climate in this area is semi-**arid** (warm, with unreliable rainfall). The land became productive farmland after **irrigation** was provided in 1912. Reliable water and available farmland attracted many people to this area.

There are two main types of farm in this area.

- Type A farms are usually about 220 hectares in size (a hectare being 10 000 square metres). Each year they grow a combination of rice, corn, wheat, vegetables and pasture, and graze beef cattle. Irrigation water is usually used.
- Type B farms are **horticulture** farms, and are usually about 20 hectares in size. They grow a combination of permanent crops that may include grapes, peaches, plums, and citrus fruit such as oranges. Many of these plants last for many years, and irrigation is always needed.

FIGURE 1 Farms in the Griffith area support businesses in the town.

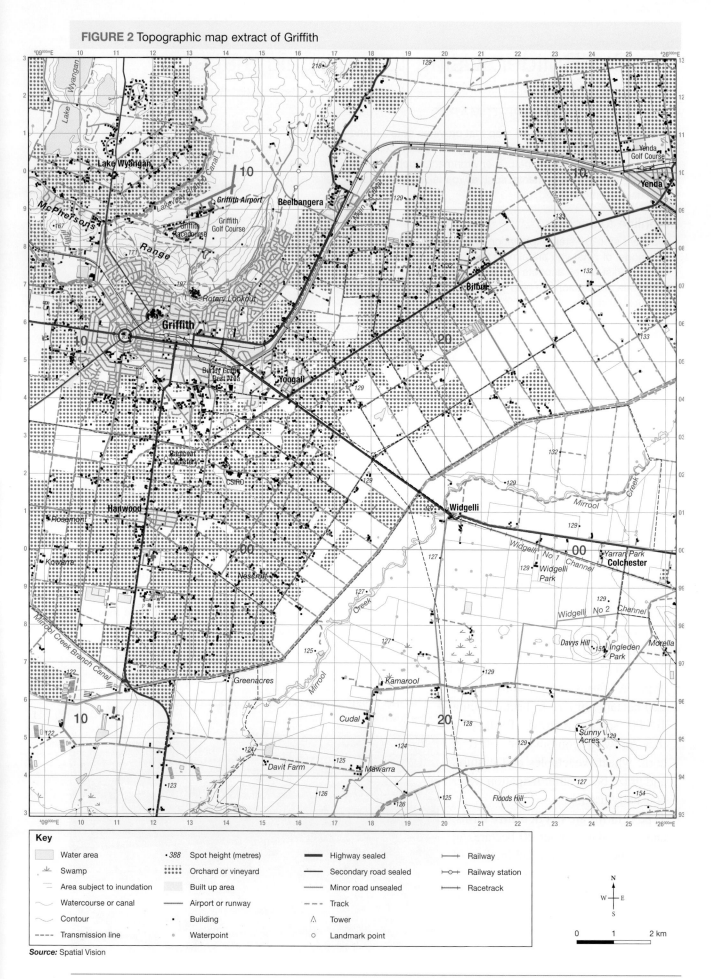

FIGURE 2 Topographic map extract of Griffith

Key

▨	Water area	•388	Spot height (metres)	▬	Highway sealed	⊢⊣	Railway
⭒	Swamp	⣿	Orchard or vineyard	—	Secondary road sealed	⊢○⊣	Railway station
⎯	Area subject to inundation	▦	Built up area	—	Minor road unsealed	⊢⊣	Racetrack
∿	Watercourse or canal	—	Airport or runway	- - -	Track		
∿	Contour	▪	Building	∧	Tower		
- - -	Transmission line	•	Waterpoint	○	Landmark point		

0   1   2 km

Source: Spatial Vision

7.7 Activities

To answer questions online and to receive **immediate feedback** and **sample responses** for every question, go to your learnON title at www.jacplus.com.au. *Note:* Question numbers may vary slightly.

For all the following activities, refer to figure 2.

Remember

1. (a) What is the main use for farmland in the area surrounding Griffith?
   (b) Sketch the symbol of this land use.
   (c) Is this an example of farming type A or type B?

Explain

2. (a) Compare the pattern made by irrigation channels and natural waterways, such as Mirrool Creek.
   (b) Why is irrigation useful in semi-arid areas? At what time of the year do you think it would be mainly used?
   (c) How can you tell from the map that it is not hilly in the areas where there is irrigation farming?

Discover

3. (a) Imagine you travelled in a southerly direction for 2.5 kilometres from the city centre. Now select one square kilometre at this location. Count the number of buildings there are in your chosen square kilometre.
   (b) Continue out from the city edge for at least another 7 kilometres. Choose another square kilometre and count the buildings in your chosen area.
   (c) Compare your results. In which area would you be closer to your neighbours?
   (d) Which one represents **intensive farming**?
4. There are many farms in the Griffith region, which means there are many people in the area to support shops, businesses, schools and cultural activities. However, in some parts of Australia, farms are very big and it is a long way to the nearest neighbours. Anna Creek, a beef cattle property in northern South Australia, is 24 000 square kilometres (2 400 000 hectares). The property is in a semi-arid region of South Australia, where vegetation is **sparse** and the nearest town for supplies is 170 kilometres away.
   (a) Use the scale to calculate the number of square kilometres covered by the map in figure 2.
   (b) How does this compare to the single farm of Anna Creek?
   (c) At which location, Anna Creek or Griffith, could you most likely satisfy each of the following wishes: to play in a sport team every week, to regularly buy clothes, to collect data about lizards, to grow a lush lawn, to safely learn to drive, to have a private airstrip?

THINK

5. Identify two natural factors and two human factors that might have influenced people to choose to live in the Griffith area.

---

learn On RESOURCES — ONLINE ONLY

Complete this digital doc: Topographic map extract — Griffith (doc-17952)

7.8 Are rural communities sustainable?

Access this subtopic at **www.jacplus.com.au**

7.9 What are 'lifestyle' places?

7.9.1 Lifestyle

In the years after 1990, the healthy state of the growing Australian and world economies meant that more and more people had jobs and were earning higher incomes. This gave them greater choice as to where and how they wanted to live, and the type of life that they wished to lead, because they had the resources (money) to allow them to choose.

For some people, 'lifestyle choice' means escaping the rush of the modern urban society by choosing a **sea change** or **tree change**. For others, it means using their new-found wealth to fulfil their wants and desires, no matter how wild. Others choose to live in inner-city areas, close to shops, cinemas, restaurants and galleries. Because of this last group of people, governments and businesses have been able to take older areas near the city centre an turn industrial zones into new activity centres where employment, residences, recreation and services can be found in the one location. Such places are in great demand by those who can afford it, particularly young professionals who want to be near the entertainment and facilities of the inner city.

For example, the Docklands development in Melbourne was designed to be the face of a new-look Melbourne: a new community identity of restaurants, entertainment and apartment living (see figure 1).

Docklands is a suburb of Melbourne, located two kilometres west of the central business district (CBD). It has a population of over 5700 people. It was planned that Docklands would be a 24-hour city for visitors and local residents. It occupies 200 hectares — an area bigger than Melbourne's central business district (which is about 175 hectares).

The local residents of this 24-hour city rely on public transport, as few have cars (see table 1). Most residents are professionals (see table 2).

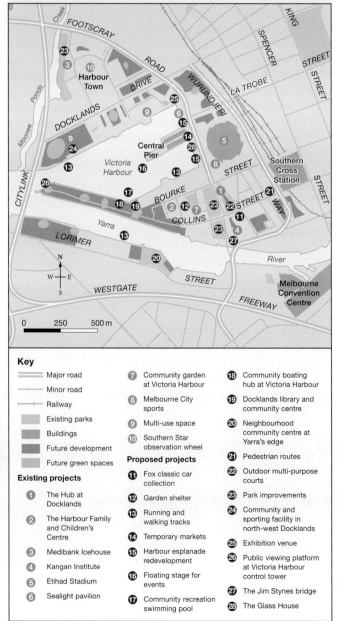

FIGURE 1 The future by 2022 and beyond: Docklands Community and Place Plan

Key

- Major road
- Minor road
- Railway
- Existing parks
- Buildings
- Future development
- Future green spaces

Existing projects

1. The Hub at Docklands
2. The Harbour Family and Children's Centre
3. Medibank Icehouse
4. Kangan Institute
5. Etihad Stadium
6. Sealight pavilion
7. Community garden at Victoria Harbour
8. Melbourne City sports
9. Multi-use space
10. Southern Star observation wheel

Proposed projects

11. Fox classic car collection
12. Garden shelter
13. Running and walking tracks
14. Temporary markets
15. Harbour esplanade redevelopment
16. Floating stage for events
17. Community recreation swimming pool
18. Community boating hub at Victoria Harbour
19. Docklands library and community centre
20. Neighbourhood community centre at Yarra's edge
21. Pedestrian routes
22. Outdoor multi-purpose courts
23. Park improvements
24. Community and sporting facility in north-west Docklands
25. Exhibition venue
26. Public viewing platform at Victoria Harbour control tower
27. The Jim Stynes bridge
28. The Glass House

Source: Spatial Vision

TABLE 1 Number of cars per dwelling

Number of registered motor vehicles	Docklands (%)	Australia (%)
None	26.8	8.6
One motor vehicle	51.9	35.8
Two motor vehicles	16.2	36.1
Three or more vehicles	3.0	16.5
Number of motor vehicles not stated	2.0	3.0

TABLE 2 Selected occupations of people

Occupation	Docklands (%)	Australia (%)
Professionals	37.7	21.3
Managers	21.5	12.9
Community and personal service workers	6.5	9.7
Technicians and trade workers	5.7	14.2
Labourers	2.1	9.4
Machine operators and drivers	1.2	6.6

FIGURE 2 Lifestyle choices for those who have the resources (a) Modern condominiums at Canary Wharf in the centre of London — a lifestyle that has arisen from the old docks (b) The historical apartments above the shops in Mala Strana in the centre of Prague, Czech Republic

FIGURE 3 The contrast between the shantytowns, or favelas, of Rio de Janeiro and their more affluent neighbours is very clear: who has the greater lifestyle choice?

FIGURE 4 Contrasting lives in India (a) The growing middle classes in modern India often purchase modern apartments, like those in this building in New Delhi. (b) For many poor Indians in urban areas, their home and way of life is on the street. Land for housing is expensive in cities such as New Delhi, and beyond the means of many people.

(a)

(b)

FIGURE 5 (a) and (b) The perfect sea change? These houses in the canal district next to Venice Beach, California, give their owners perfect peace and tranquility. (c) … or do they? (d) The beachfront houses on Venice Beach (e) Condominiums on the cliffs of the Pacific Ocean, in the well-off suburb of Santa Monica, California (f) Many homeless people in California opt to live on Venice Beach because of the climate.

(a)

(b)

(continued)

FIGURE 5 continued

(c)

(d)

(e)

(f)

7.9 Activities

To answer questions online and to receive **immediate feedback** and **sample responses** for every question, go to your learnON title at www.jacplus.com.au. *Note:* Question numbers may vary slightly.

Explain

1. Study figures 2, 3, 4 and 5. Identify the features in the photographs that indicate that the people living in these *places* have the resources to choose to live in such *environments*.
2. Create a cartoon that summarises the differences between a sea *change* and a tree *change*.
3. Compare the images in figures 3 and 4. What similarities and differences can you see in the 'lifestyle' choices of these residents of Rio de Janeiro and New Delhi? How much real power do they have to decide where they will live?
4. Refer to table 1.
    (a) Identify two key facts the table reveals about the number of cars Docklands residents have compared to the average for other Australian suburbs.
    (b) Think of a reason to explain this.
5. Refer to table 2.
    (a) Identify two key facts the table reveals about Docklands residents compared to the average for other Australian suburbs.
    (b) Think of a reason to explain this.
6. Refer to figure 1. Which features (existing or proposed) would make Docklands a 24-hour suburb?

Discover

7. Use online resources to investigate the current state of the Docklands development in Melbourne. Identify the planned zones and features that are in operation and those that have been changed, removed or not built.
8. Based on your research, do you believe that the Docklands *space* provides an enjoyable lifestyle for the people who live there? Give reasons for your answer.

Predict

9. Study the lifestyle *environment* in figure 5. From the evidence in the images, identify and list any possible *changes* (natural or human) that might affect the liveability of Venice Beach and Santa Monica.
10. Would you like to live in a *place* like Venice Beach? Give reasons for your answer.

Think

11. In groups of four, use a graphics software program to create a concept wheel that explores the meaning of the word lifestyle. Display all the wheels in a class electronic presentation.
12. As a class, compare and discuss the lifestyle wheels that you produced in question 11. Is there a common view of lifestyle that is representative of the whole class?

7.10 Where is my place?

7.10.1 My place

What is your **neighbourhood** or local place like? All of us live in a community, and these are often centred around the place where we live, go to school or work.

Teenagers have different types of local places that have special meaning for them, each one at a different scale: their bedroom, home and neighbourhood.

When you live in a neighbourhood, you become familiar with all the things that help to create the character of the place. Sometimes a neighbourhood is made up of people who have similar interests and beliefs, whether these be cultural, sporting, environmental or job-related. Other neighbourhoods have a mixture of people from different backgrounds, creating a vibrant, multicultural community identity. The fact that Australian neighbourhoods can be so different is what makes Australia such an interesting place to live in.

Neighbourhoods have always existed in Australia. The 'country' that is special to the many Aboriginal and Torres Strait Islander nations is often based on language. For instance, the Kulin Nation consists of five Victorian Aboriginal communities who lived in what is now the Melbourne region before Europeans invaded. Each community spoke its own language and controlled a region that had definite boundaries (see figure 1). Within each community, there were different dialects that overlapped. These dialects were spoken by different clans — groups of related families. Thus, these nations saw, and often still see, their neighbourhood as the region in which people spoke the same language and had the same customs, such as marriage rituals. People were, and are, socially connected.

Because nearly 90 per cent of Australians live in towns and cities, most people are likely to live in a street that is part of a suburb, town or city, and which itself is part of a state or territory. On the other hand, there are Australians who do not live in urban areas, but still live in their own communities that are just as distinctive as neighbourhoods in towns and cities. How can we describe where our local place is and what it is like? Sometimes, people try to use words to do this, but it is not an easy task. Geographers have no such trouble, however; they can use maps.

FIGURE 1 The places belonging to Indigenous Australian peoples

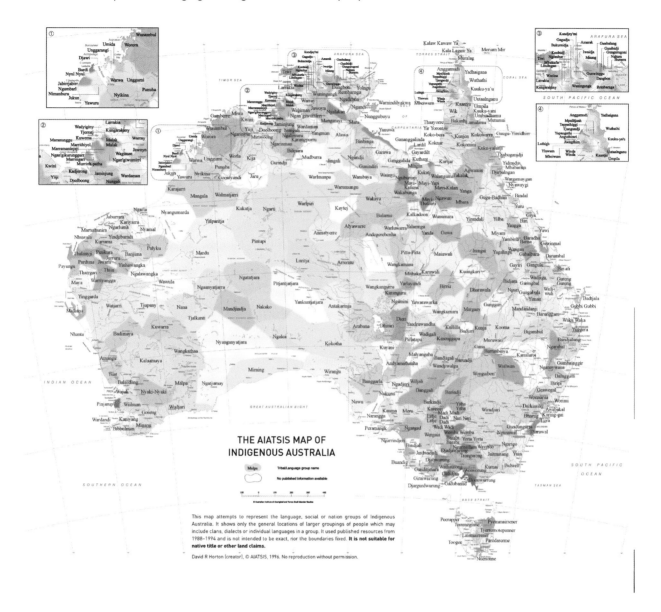

FIGURE 2 Mental map of Jayden's local place (a) by Jayden and (b) by Annette, Jayden's mother

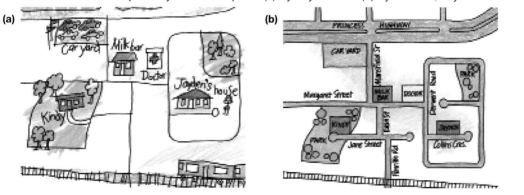

7.10 Activities

To answer questions online and to receive **immediate feedback** and **sample responses** for every question, go to your learnON title at www.jacplus.com.au. *Note:* Question numbers may vary slightly.

Think

1. Figure 2 shows two mental maps of the same neighbourhood *place*. One is drawn by Jayden, a Year 7 boy, and the other is drawn by his mother. Compare the two maps by drawing up a table like the one below and filling in the spaces.

	Features that are different	Features that are similar
Land use		
Transport		
Street layout		
Relative sizes		
Names of places		
Other		

Explain

2. Suggest reasons to explain the major similarities and differences between the maps drawn by Jayden and his mother. Think about factors such as age, duties during the day, transport and friendships.

Discover

3. Create a mental map of your neighbourhood or local *place*. Locate your house in the centre of the sheet and work outwards from there. The map should be as detailed as possible. Include features such as:
   - streets and their names
   - houses of friends or family
   - shops, parks, trees, post boxes, telephone poles, pedestrian crossings, railway lines and stations
   - anything you can remember, but the map must be drawn from memory.
     Present the map using geographical rules (BOLTSS). Since you are not drawing the map to a *scale*, write 'Not to scale' in the correct position. Remember to use conventional colours and symbols as far as possible. Compare your mental map to an actual map of your neighbourhood.
   (a) In what ways was your map accurate?
   (b) Which features did you not mark on your map?
   (c) Which parts of your neighbourhood did you know well and which did you not know well?
   (d) Think of reasons to explain your answers to part (c).

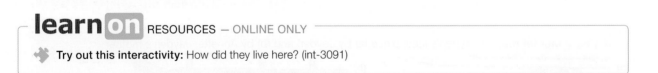

learn on RESOURCES — ONLINE ONLY

Try out this interactivity: How did they live here? (int-3091)

7.11 Where would you like to live?

7.11.1 What makes this place so liveable?

Where is your favourite place in Australia? Have you been to a holiday paradise, one that you think would be the perfect place to live? Is the climate perfect, the scenery spectacular? Is it safe, fun and the place for adventure? Is this place in a city, in the **wilderness** or in the next street? Is it paradise because your friends or family live there or because of the natural or **built environment**?

Among the most popular and beautiful tourist destinations in Australia are the Great Barrier Reef, Uluru, Melbourne, Sydney, the Gold Coast, the Great Ocean Road, Monkey Mia, Kakadu, the Tasmanian

Wilderness, the Blue Mountains, Port Arthur, Byron Bay, Kangaroo Island and Ningaloo Reef. Many of these places have unique landscapes, located within naturally stunning environments. Four of these are predominantly built environments: Sydney, Melbourne, the Gold Coast and Port Arthur. The remaining 11 places are best known for their natural, often remote, and almost wilderness environments.

Some of these wonderful places are found in or close to cities and large towns; some have significant local populations; and some are quite remote. They are all places that attract large numbers of visitors every year. People come to see or experience an aspect of the local environment that brings them pleasure. These places are often perfect for a holiday but they may also be a place to live. Is it mostly the excitement of a big city, natural beauty, or some other factor that makes you decide which place is the most liveable?

FIGURE 1 Five of Australia's most popular places for tourists

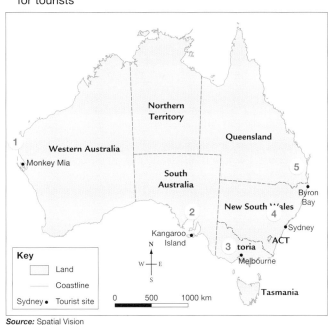

Source: Spatial Vision

1 Monkey Mia is an environment where you can experience natural wildlife by interacting with dolphins. Monkey Mia is located in Shark Bay on the coast of Western Australia, 850 kilometres north of Perth. For over 40 years, a small pod of dolphins has come ashore to connect with beachgoers. The Department of Environment and Conservation provides staff who supervise the feeding of fish to these dolphins each day. It is an unusual opportunity for people to see wild dolphins up close, quite near to the shore. Monkey Mia is a place of great natural beauty without a huge tourist resort attached. Most visitors camp. It is an important stop on the around-Australia tourist trail. Fewer than 800 residents currently live near the Monkey Mia Resort.

2. Kangaroo Island is a place of natural beauty. It is Australia's third largest island, found about 160 kilometres south of Adelaide. It is a wildlife lover's paradise, being home to many native Australian animals in their natural habitats, including koalas, kangaroos, seals and penguins. It has remote, unspoiled beaches and interesting rocky outcrops. Although first settled in the late 1830s, its present population of over 4200 is the highest it has ever been. It was originally settled as a fishing and farming community but today is better known as a tourist destination.

3. Melbourne is the second largest and most **liveable city** in Australia (2011–2015, the *Economist* magazine). It is the capital of Victoria and home to about 4.4 million residents. It is an attractive destination for tourists, who enjoy visiting its major sporting and cultural events, shops, restaurants and theatres. Melbourne is located beside Port Phillip Bay and on the Yarra River. It is not a city known for its beautiful natural environment, but it has become known for its distinctive laneways, bars and café culture.

4 Sydney is a built environment in a beautiful setting and is Australia's largest and oldest city. It is often called the 'Harbour City'. Sydney is popular with both domestic and international tourists and is home to 4.8 million residents. It has many attractions, including restaurants, beautiful beaches, theatres, galleries and iconic landmarks. It has a beautiful natural environment with varied experiences provided by the built environment. This makes it an extremely popular destination for everyone.

5 Byron Bay is a beachside town in northern New South Wales, located 160 kilometres south of Brisbane. Byron Bay is a very relaxed place with a local community that includes many artists and retired hippies. It is an important surfing place, with easy access to offshore reefs and stunning beaches. It has become a popular place for 'schoolies' end-of-year celebrations. Byron Bay has a population of about 30 000 people, who rely heavily on tourism and agriculture for their income.

7.11 Activities

To answer questions online and to receive **immediate feedback** and **sample responses** for every question, go to your learnON title at www.jacplus.com.au. *Note:* Question numbers may vary slightly.

Explain

1. Brainstorm a list of the features to describe your most liveable *place*.
2. Is there a *place* in Australia that you have been to or heard about that is your most liveable *place*? Locate this *place* on a map and use images from the internet or magazines to explain how it fulfills your list of features from question 1.

Discover

3. Is your most liveable *place* in a natural or a built *environment* or a mixture of the two?
4. What would be the advantages and disadvantages of living in your most liveable *place*?
5. After reading the paragraphs numbered 1, 2, 3, 4 and 5, which of these *places* is most similar to your most liveable *place*? Explain your answer.

Predict

6. If you were looking for a *place* to retire, which of the *places* in figure 1 would be best and why?
7. If you wished to work as a national park ranger, which of the *places* in figure 1 would be best and why?
8. If you were planning a career in the theatre, which of the *places* in figure 1 would be best and why?
9. If you wished to live in a relaxed coastal *environment* close to a capital city, which of the *places* in figure 1 would be best and why?

Think

10. Design a map of your most liveable *place*. Consider the natural and built *environments*; distance to a city, services, job and recreational opportunities; climate; and lifestyle. Annotate your map to explain why this is where you would like to live. Use the **Nothing like Australia** weblink in the Resources tab to help find your ideal location.

 RESOURCES — ONLINE ONLY

 Explore more with this weblink: Nothing like Australia

7.12 SkillBuilder: Creating a concept diagram

online only

WHAT IS A CONCEPT DIAGRAM?

A concept diagram, sometimes mistakenly called a concept map, is a graphical tool that shows links between ideas, or concepts. Concept diagrams organise links into different levels.

Concept diagrams enable you to organise your ideas and communicate them to others.

Go online to access:

- a clear step-by-step explanation to help you master the skill
- a model of what you are aiming for
- a checklist of key aspects of the skill
- a series of questions to help you apply the skill and to check your understanding.

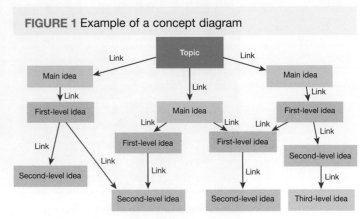

FIGURE 1 Example of a concept diagram

7.13 Review

7.13.1 Review

The Review section contains a range of different questions and activities to help you revise and recall what you have learned, especially prior to a topic test.

7.13.2 Reflect

The Reflect section provides you with an opportunity to apply and extend your learning.

Access this subtopic at **www.jacplus.com.au**

TOPIC 8
People and places

8.1 Overview

Numerous **videos** and **interactivities** are embedded just where you need them, at the point of learning, in your learnON title at www.jacplus.com.au. They will help you to learn the content and concepts covered in this topic.

8.1.1 Introduction

We all live in different places. Places are important to people, whether they are rural or urban, remote or central, permanent or temporary. But no two places are alike; they differ in aspects such as their appearance, size and features. In your mind's eye, try to picture the similarities and the differences between places such as a country town, a popular tourist destination, a remote village overseas, a scientific base in Antarctica, an Indigenous community and a mining town. You may think of others to add to this list.

Starter questions

1. Brainstorm, as a class or individually, a list of other types of *places* in which people live around the world.
2. Choose three examples of *places* on your list that have been strongly influenced by the quality of the *environment*. Explain how the *place* is influenced by the *environment*.
3. Other than *environmental* qualities, what influences the characteristics of a *place*?
4. How true is it that 'no two places are the same'? Use images in this section, or *places* you know, to support your opinion.

The places people live in around the world have many different characteristics and features.

8.2 Do most Australians live near water?

8.2.1 Historic settlements and transport networks

Does the availability of rainfall explain why Australians currently live where they do? Could it also be warmer weather, good soils, and access to the coast, mineral resources, people and flat land?

Fresh water availability and a coastal location have always influenced the places where Australia's inhabitants live, because our continent is huge, dry and isolated. Unlike Indigenous Australian peoples, the people who occupied Australia after 1788 have not been tied to a particular place in Australia by their understanding of '**country**'. The liveability of places can be influenced by their access to resources.

Australia's first large European colonies were all built close to rivers and harbours, which provided safe anchorage for their sailing ships. The early colonisers and convicts, along with their possessions and food, were all transported from Europe by ship. These colonisers were quite wary of the unknown inland of the continent, so they clung to the coast and relied on sea transport. The sea allowed goods to be imported and exported; it provided fishing, whaling and sealing; and it brought cooler weather due to onshore breezes. Rivers, such as the Yarra (seen in figure 1), provided water for household, industrial and agricultural use, as well as a safe port for passengers and cargo.

Sydney, Melbourne, Brisbane, Hobart, Adelaide and Perth were generally the first settlements in each colony. They have maintained their importance as their state's largest city and its centre of government, transport and commerce. Big cities are great places to live because they provide many opportunities for work, education, healthcare and recreation.

FIGURE 1 The Yarra River, Melbourne, 1864

Land use

Beginning in 1788, Europeans and subsequent migrants have changed the natural environment by clearing vegetation and using the land for building cities and farms and creating the services people need.

FIGURE 2 The major land use regions of Australia

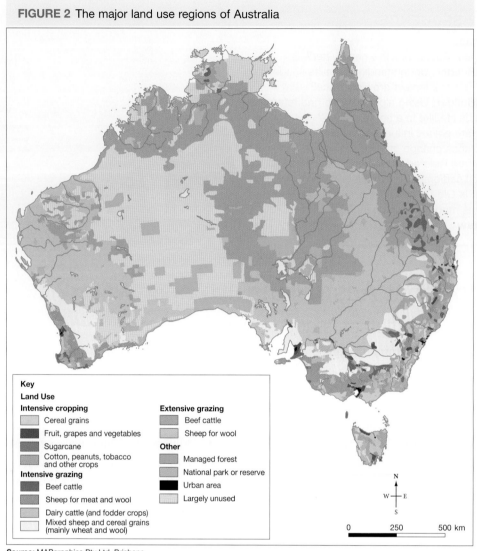

Key

Land Use

Intensive cropping
- Cereal grains
- Fruit, grapes and vegetables
- Sugarcane
- Cotton, peanuts, tobacco and other crops

Intensive grazing
- Beef cattle
- Sheep for meat and wool
- Dairy cattle (and fodder crops)
- Mixed sheep and cereal grains (mainly wheat and wool)

Extensive grazing
- Beef cattle
- Sheep for wool

Other
- Managed forest
- National park or reserve
- Urban area
- Largely unused

N
W—E
S

0   250   500 km

Source: MAPgraphics Pty Ltd, Brisbane

Examine figure 2 and try to work out the difference between **intensive** and **extensive** land use regions. The intensive regions are mostly located closer to the coast than extensive regions, and they are smaller areas. Compare this land use map to a rainfall map in your atlas. Does more rain tend to fall in the intensive or the extensive land use regions?

In places where rainfall is low or unreliable, such as central Australia, grass growth is seasonal. This semi-arid interior does not support the variety of land uses that are possible closer to the coast. Over 80 per cent of Australia's population lives within 500 kilometres of the coast between Brisbane and Adelaide; some are on farms but most are in regional towns and capital cities.

Jobs

People need to work in order to provide for their basic needs. Figure 3 shows that the majority of jobs in 2015 were in the sectors of services (health, finance, education and administration), retail trade and construction. Most of these job opportunities are available in major cities and regional centres, which are generally close to the coast and in areas of higher rainfall.

There are few job opportunities in dry and remote regions of Australia. Agriculture and mining provide some chances for employment in these places. Some jobs are also available in manufacturing (processing the output of farms and mines) and in services, such as healthcare and tourism. According to the Australian Bureau of Statistics, 25 per cent of new jobs created in the next 10 years will be in the services sector, especially healthcare, but these will mostly be in the cities. Mining jobs in remote places are declining due to the reduced international price for minerals and mine closures and very few new mines opening. Remote mining sites have skills shortages that require employers to pay very high wages to attract workers.

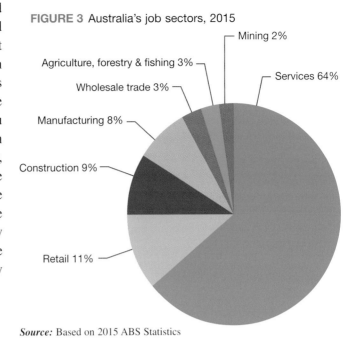

FIGURE 3 Australia's job sectors, 2015

- Mining 2%
- Agriculture, forestry & fishing 3%
- Wholesale trade 3%
- Services 64%
- Manufacturing 8%
- Construction 9%
- Retail 11%

Source: Based on 2015 ABS Statistics

8.2 Activities

To answer questions online and to receive **immediate feedback** and **sample responses** for every question, go to your learnON title at www.jacplus.com.au. *Note:* Question numbers may vary slightly.

Remember

1. List four reasons why the state capital cities were settled.
2. In Australia, are more workers employed in mining or in manufacturing?

Explain

3. What is the difference between intensive and extensive land use? Give examples of each type.
4. Locate and study a rainfall map of Australia (*Jacaranda Atlas Eighth Edition*, page 40) and the land use map of Australia (figure 2). Approximately how many millimetres of annual rainfall do areas of (a) intensive farming receive and (b) extensive farming receive?

Discover

5. Are there regions of intensive agriculture found in places with much less rainfall? Where are they? Investigate what technology these farmers might use to increase the available water for their crops?
6. 'Access to water improves the liveablility of a place.' Write a paragraph that agrees with this statement and another that disagrees with this statement.

Predict

7. Design a cartoon that comments on future environmental conflict between water for everyday living in cities and water for farming in Australia. Remember that only about two per cent of employment in Australia is on farms.

Think

8. A doctor may be offered very high wages to work in a remote, dry place in north-western Australia. What economic, *environmental* and social measures might they use to decide between the liveability of this *place* and their current workplace in Melbourne?

8.3 SkillBuilder: Understanding satellite images

WHAT ARE SATELLITE IMAGES?

Satellite images are images that show parts of our planet from satellites in space and transmitted to stations on Earth. Satellite images help geographers observe a much larger area of the Earth's surface than photographs taken from an aircraft.

Go online to access:

- a clear step-by-step explanation to help you master the skill
- a model of what you are aiming for
- a checklist of key aspects of the skill
- a series of questions to help you apply the skill and to check your understanding.

FIGURE 1 A false-colour satellite image of the Mt Lofty Ranges, South Australia

RESOURCES — ONLINE ONLY

Watch this eLesson: Understanding satellite images (eles-1643)

Try out this interactivity: Understanding satellite images (int-3139)

8.4 Where is the fastest growing place?

Access this subtopic at **www.jacplus.com.au**

8.5 SkillBuilder: Using alphanumeric grid references

WHAT ARE ALPHANUMERIC GRID REFERENCES?

Alphanumeric grid references are a combination of letters and numbers that help us locate specific positions on a map. Letters and numbers are placed alongside the gridlines, just outside a map. The grid, letters and numbers allow you to pinpoint a place or feature by stating its alphanumeric grid reference.

Go online to access:

- a clear step-by-step explanation to help you master the skill
- a model of what you are aiming for
- a checklist of key aspects of the skill
- a series of questions to help you apply the skill and to check your understanding.

FIGURE 1 A map of Canberra and its suburbs with an overlaid alphanumeric grid

8.6 What is life like in a country town?

8.6.1 The attraction of the country

Country towns come in all shapes and sizes. They can be small centres with a post office and general store or they can be substantial towns. Because most of Australia's population and businesses are concentrated in the capital cities, even people who live in quite large towns outside the capital cities see themselves as living in the country.

Even though most Australians live in large urban centres, the rural or country regions are very important because this is where food is grown, water is sourced, **natural resources** are extracted and ecosystems can flourish. Many Australians travel to country places for holidays and many dream of moving to the country. The attractions of country places include cheaper housing, less traffic and a greater sense of safety.

8.6.2 Demography

The **demographic** characteristics of country places are influenced by location and activities in the surrounding area.

For instance, Leongatha is located on the South Gippsland Highway, 135 kilometres south-east of Melbourne, Victoria. Reliable rainfall and good soil make the area one of the most productive in Victoria. Dairy farming is the main type of farming, and the milk-processing factory is the largest single employer in town.

Another town, Coleraine, is located on the Glenelg Highway, 350 kilometres west of Melbourne. The farms are generally large. Sheep and cattle grazing are the main types of farming, and there is no major business in the town.

TABLE 1 Predicted population for selected Victorian places

Local government area	2011		2031	
Municipality	% aged under 20	% aged over 65	% aged under 20	% aged over 65
Melbourne (urban)	24.3	23.8	13.0	17.1
South Gippsland Shire (rural includes Leongatha)	24.8	21.9	19.5	28.1
Southern Grampians Shire (rural includes Coleraine)	25.8	21.6	19.6	30.3

A sense of belonging

In the country there are opportunities to be involved in a wide range of community activities. Activities might have an economic focus (such as fundraising), an environmental focus (such as Landcare) or a social focus (such as youth groups). A common outcome of all activities is the way they contribute to a sense of connectedness or belonging.

It is common for even small towns to provide a range of sports. Sports provided in a town as large as Leongatha might include Australian Rules football, cricket, little athletics, tennis, equestrian events, bowls, fishing, cycling, croquet, skateboarding, golf, swimming, basketball, netball, table tennis, badminton, karate, gymnastics, squash and taekwondo. Also available are cultural activities and entertainment, such as films, brass bands, guides, scouts, art galleries, dancing and theatre groups.

FIGURE 1 Growing up in a country town might mean …

- Fewer restrictions on your time
- A greater understanding of the natural world (e.g. plants, animals, weather) and appreciation of the importance of the quality of the environment
- A sense of safety (e.g. cars and houses not always locked)
- A small group of people your age
- Little public transport; reliance on car travel
- A volunteer fire brigade
- Clean air
- An understanding of Aboriginal traditions
- Greater likelihood of seeing native animals
- A wider range of pets
- Travelling to school by bus
- Leaving home for tertiary study
- Opportunities for physical activity
- Uncrowded roads
- Plenty of space to move about in
- A limited range of specialists, shops and employment
- Quiet walking and riding trails and fishing spots
- Knowing many people in the community and many people in the community knowing you
- Community reliance on surrounding resources (e.g. farming, mining, fishing or the environment that attracts tourists)
- Fewer films and less live music

FIGURE 2 Lawn bowls is a popular sport in small towns.

8.6 Activities

To answer questions online and to receive **immediate feedback** and **sample responses** for every question, go to your learnON title at www.jacplus.com.au. *Note:* Question numbers may vary slightly.

Remember

1. In which direction would you travel from Leongatha to reach Melbourne? In which direction would you travel from Coleraine to get to Melbourne?
2. Provide three reasons why country *places* are important.

Explain

3. Classify each of the characteristics in figure 1 as economic, social or *environmental*.
4. Identify the characteristics in figure 1 that are attractive to you. Are most of these characteristics social, economic or *environmental*?
5. What are three opportunities in your community for young people to feel socially connected?

Discover

6. Study the data in table 1.
   (a) The percentage of the population under 20 is predicted to _____ in all municipalities.
   (b) True or false? Country regions are predicted to have a big increase in the percentage of population over 65.
   (c) What is the evidence to support your choice for part (b)?
   (d) Apart from the age of the population, suggest another difference in the demographic between country towns and big cities.
7. Indigenous Australians have always had a sense of belonging to a *place*, and elements of the natural *environment* are honoured on their flags. Find out what the Aboriginal flag and the Torres

Strait Islander flag look like. What aspects of the natural *environment* are represented on each of these flags?

8. As a class or group, brainstorm the meaning of the term *country life*. Create a list of words that describe your agreed view. Choose a suitable strategy to convey your representation of rural living.

9. (a) Create a list of Australian country towns that you have heard of.
   (b) Collaborate with others to map the location of each of these towns.
   (c) For each town, record its latitude, distance and direction from the state capital city.
   (d) Complete the following generalisation about the known towns:
       'Most of the towns we know are located _____.'

Think

10. In Australia, the trend is for people to move away from the country to the major cities. Suggest three reasons why you think this happens.

RESOURCES — ONLINE ONLY

Try out this interactivity: Country town services (int-3092)

8.7 How are places influenced by seasons?

Access this subtopic at **www.jacplus.com.au**

8.8 Do places change over time?

8.8.1 On the move

A town will change over time if the factors influencing people's decision making about living there also change. Change may be due to government plans, the perception of the natural environment, the economic activities that are carried out in the place and access to resources and other places.

The original buildings in Tallangatta, in north-east Victoria, about 40 kilometres from Albury and Wodonga, can be seen only when the water level in Lake Hume is very low. The current town was moved from its original location in 1956. Houses were lifted onto trucks (with parts of the buildings often falling off during the journey) and moved about eight kilometres (see figure 1). The original site, in a valley beside the Mitta Mitta River, was flooded when the size of Lake Hume was increased.

8.8.2 Town closed

In 1917, it was decided that a town was needed on the dry and very warm Nullarbor Plain to provide services for the Indian Pacific railway (see figure 2). With a population of 300, the town of Cook was once big enough to have a school, hospital, shop and accommodation for train drivers. When the railways were privatised in 1997, the town was closed. The population now stands at four, and the one shop is open only when the Indian Pacific is in town.

FIGURE 1 A Tallangatta house being moved to the new town site

FIGURE 2 The location of Cook

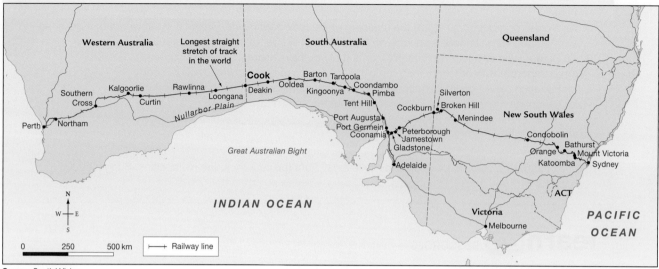

Source: Spatial Vision

8.8.3 Access to resources

Resource depletion

Silverton, 25 kilometres north-west of Broken Hill, was once home to 3000 people who mainly worked in mining (see figure 3). Most people left, often taking their homes, when richer mines opened at Broken Hill. The current population is 50, and the town is now visited by many tourists. The town and its semi-arid surroundings have been used as the setting in many films, such as *Mad Max 2, Dirty Deeds, Mission Impossible 2* and *The Adventures of Priscilla, Queen of the Desert*.

Resource discovery

Karratha is a hot, dry place 1600 kilometres north of Perth. It was founded in the 1960s for workers on the growing iron ore mines in the Pilbara region. In the 1980s the development of the natural gas industry encouraged further growth. The town currently supports about 28 000 people and is expected to support up to 40 000 by 2030.

8.8.4 Sea change

Margaret River, 270 kilometres south of Perth, has become popular because it offers a rural lifestyle and is accessible to the capital city. People who move from the city to the coast are said to have made a 'sea change'. Those who move to an inland location are said to have made a 'tree change'.

FIGURE 3 The hotel in Silverton and the surrounding landscape have been used in films such as *Mad Max 2*.

FIGURE 4 The planned town of Karratha

Change over time

Many people now recognise that the Margaret River region has many attractions, such as beaches, waterways, caves, wineries, national parks and mild weather that suits farming and tourism.

However, what people have thought about the region has changed over time. Before 1830, the Noongar people, including the Wardandi Nation, valued the natural characteristics (such as flora, fauna, weather, sea and rivers) and made few changes to the natural environment. In 1830, white settlers arrived to cut down trees and sell timber. In 1950, they began using the cleared areas for dairy cattle and beef cattle.

Tourists also began to value and visit the region's natural features, such as beaches, rivers and caves. By 1970, people were moving from the city to enjoy the quiet country atmosphere and, by 1990, the area had become popular as a sea change destination.

FIGURE 5 The Bussell Highway — the main street in Margaret River — 1991

TABLE 1 Population change in Margaret River

Year	Population of town
2001	3627
2006	4415
2011	5314
2016*	6700
2021*	8500

* predicted population

TABLE 2 Origin of people who moved to Margaret River, 2006–2011

Previous place of residence	Number
New South Wales	83
Victoria	35
Queensland	44
South Australia	30
Western Australia	1004
Tasmania	13
Northern Territory	20
Overseas	365

8.8.5 Tourism

Port Douglas, 60 kilometres north of Cairns, was a busy port in the 1870s and had a population of more than 10 000. The mining that had attracted people to this hot, wet area did not last. By the 1960s, the population was only 100. In the 1980s, road and air access to the town improved. People were prepared to travel long distances from within Australia and from overseas to enjoy the warm weather, stunning beaches and the World Heritage areas of the Great Barrier Reef and Daintree rainforest. The permanent population is now about 1300. During the peak holiday season (May to November) there are at least four times this number of people in Port Douglas.

FIGURE 6 Port Douglas in 1971, before the tourist boom

FIGURE 7 Port Douglas in 2009

Dickson Inlet

8.8.6 Change in the future

Even in a small state like Victoria, predicted population growth varies across the state. Towns relying on big farms are predicted to lose population. The use of machinery and the closure of processing plants have reduced employment opportunities. Towns in regions very close to Melbourne are predicted to grow. People who live in these places still have access to jobs and entertainment in Melbourne even though they live in regional Victoria. More people means there is a need for more businesses.

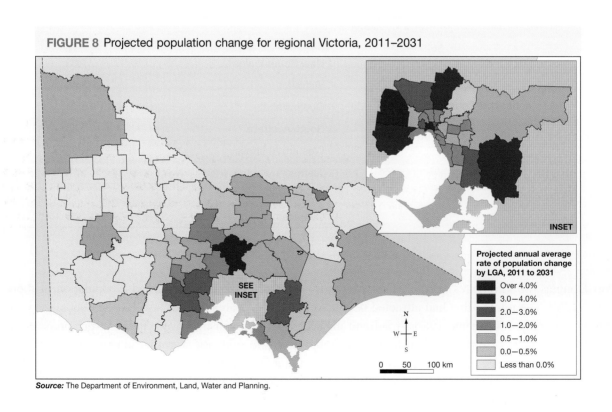

FIGURE 8 Projected population change for regional Victoria, 2011–2031

INSET

SEE INSET

Projected annual average rate of population change by LGA, 2011 to 2031

- Over 4.0%
- 3.0–4.0%
- 2.0–3.0%
- 1.0–2.0%
- 0.5–1.0%
- 0.0–0.5%
- Less than 0.0%

0   50   100 km

Source: The Department of Environment, Land, Water and Planning.

8.8 Activities

To answer questions online and to receive **immediate feedback** and **sample responses** for every question, go to your learnON title at www.jacplus.com.au. *Note:* Question numbers may vary slightly.

Remember

1. Which water storage drowned old Tallangatta? How far was the town moved?
2. In which region of Western Australia is Karratha?
3. What now draws people to Silverton?
4. What is the population of Port Douglas in the peak holiday season?

Explain

5. (a) How many states does the Indian Pacific travel through?
   (b) Why do you think the train is called the Indian Pacific?
   (c) In which general direction does the train travel from Sydney to Perth?
6. Read the description of the *change* over time for Margaret River. Create a timeline to show the *changing* view of Margaret River.
7. (a) Refer to tables 1 and 2 and calculate the population increase between each census. In which time period was the greatest population increase in Margaret River?
   (b) What are the three main *places* new residents came from to settle in Margaret River between 2006 and 2011?
8. Refer to your atlas or find a climate graph and suggest why May to November is the peak holiday season in Port Douglas.
9. Refer to the map in figure 8.
   (a) Describe the location of the areas predicted to grow by more than 3 per cent. For, example, are they inland or by the coast? Are they in the north, south, east or west of the state? Are they clustered together or spread out? Are they close to Melbourne?
   (b) What will happen to towns in the palest pink regions?
   (c) Estimate the proportion of Victoria that is predicted to increase its population and the proportion that is predicted to decrease its population.
10. Find maps of Victoria that provide information about landform and climate. Refer to your maps and figure 8 to complete the following.
    (a) Think about landform and population *change*. Are most areas of declining population in *places* that are not mountainous? Are most areas of increasing population on the coast side of the mountains?
    (b) Think about climate and population *change*. Are most of the highest growth population areas in *places* where rainfall is over 600 millimetres per year? Are most areas of declining population in *places* where the rainfall is lower?
    (c) What might be reasons for your findings in (a) and (b)?

Discover

11. (a) Draw a sketch map of natural features at Port Douglas in 1971. Show the ocean, promontory, Dickson Inlet, flat land and hills. Add the settlement features (such as housing, roads and marina).
    (b) Using another colour or an overlay, show the settlement features for 2009.
    (c) Annotate your map to describe the *changes* that have occurred and the *changes* you think will happen in the next 10 years.
12. Provide another example of a town in Victoria that has *changed*. Clearly describe its geographic characteristics:
    • location — state, latitude, distance and direction from your school
    • surrounding landform
    • climate
    • significant physical features (e.g. mountain, river, ocean or lake).
    Collect images and data to show that *change* has occurred. Explain why the town has *changed*. Possible sources of information include your atlas, the local council, local newspaper and the Australian Bureau of Statistics.

8.9 What are isolated settlements like?

8.9.1 Way up north

Life in some towns is strongly influenced by geography. It is now possible to live in extreme conditions and be socially connected while enjoying access to a wide range of goods, services and community activities.

Dawson, with a population of 2000, is the second largest city in the Canadian state of Yukon. At the latitude of 64°N, it is only about 360 kilometres from the Arctic Circle.

Dawson is a long way from neighbouring towns. It is 770 kilometres from the next town to its north (Inuvik) and 810 kilometres to Anchorage, the largest town in Alaska.

8.9.2 Climate

TABLE 1 Average hours of sunlight at Dawson, Yukon and Longreach, Queensland

Month	Dawson, latitude 64°N	Longreach, latitude 23°S
January	4	13.5
February	6.9	13
March	10.2	12.5
April	13.7	12
May	17.1	11
June	20.6	11
July	21.4	11
August	18.1	11
September	14.6	12
October	11.2	12.5
November	7.8	13
December	4.7	13.5

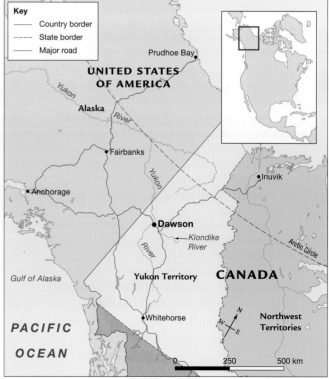

FIGURE 1 Location of Dawson, Yukon, Canada

Source: Spatial Vision

8.9.3 Settlement

The site of Dawson, on the junction of the Yukon and Klondike rivers, was always an important hunting and meeting site for indigenous peoples, who still live in the area. The harsh climate deterred white colonisers from the area until gold was discovered in 1896. The population grew by thousands every week, quickly reaching 30 000.

However, most people did not stay long. Living conditions were very poor: people lived in tents and huts; they had no power, water or sewerage; and crops would not grow in the low temperatures. The **permafrost** also made it difficult to dig the foundations for buildings. Furthermore, it turned out that the gold was not accessible: it was in gravel that was frozen for most of the year.

Today

Today the town has infrastructure, and housing is solid and heated. The community of Dawson operates year round, with the climate having a strong influence on activities. A road to Alaska has encouraged tourism, and new mining techniques have increased goldmining.

The weather conditions mean that some jobs, such as road-making and building, cannot be done in the winter. Some of those employed in these industries move south in the winter, either for work or holidays.

The town of Dawson is no longer completely isolated in winter. Internet and phone connections allow communication; snow-ploughing and spreading of sand keep the roads open most of the time; and planes fly in and out all year.

The weather influences many aspects of social life in Dawson. The 'Thaw di Gras' carnival celebrates the end of winter, and music festivals and a kayak marathon take place in summer. Winter is the time of snowmobile treks, skiing and the famous Yukon Quest dog-sled race from Whitehorse (Yukon) to Fairbanks (Alaska). During the race, dogs and mushers (sled drivers) need protection from the cold, but if the weather warms to minus 4 °C, it becomes too hot for the dogs to travel during the day.

FIGURE 2 Climate graph, Dawson

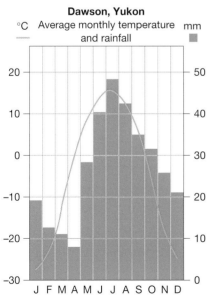

FIGURE 3 It is so cold in winter that the Yukon River freezes over, and the ice is thick enough to support fully loaded trucks.

8.9 Activities

To answer questions online and to receive **immediate feedback** and **sample responses** for every question, go to your learnON title at www.jacplus.com.au. *Note:* Question numbers may vary slightly.

Remember

1. Refer to the map in figure 1. How far is Dawson from Whitehorse, the state capital?
2. (a) What influenced indigenous peoples to live in the Dawson area?
   (b) Why did people move to Dawson in 1897?
   (c) What influences people to live in Dawson nowadays?

Explain

3. List at least four ways in which the natural *environment* has influenced the settlement of this area.
4. What factors have improved liveability in Dawson?
5. (a) Construct a multiple bar or column graph to show the average number of hours of sunlight at Dawson and Longreach.
   (b) On your graph, use colour, shading or symbols to show summer and winter at each location.
   (c) Write one statement that is true for both Dawson and Longreach about their hours of sunlight.
   (d) Imagine you live in Dawson, where the school year is from August to June. In which months would you arrive or leave in the dark?
6. Activities in Dawson include volleyball, curling, film festivals, craft, dances, motorcycle riding, gold panning, softball, writing workshops and outhouse races. Which of these would occur only in summer? Which would occur only in winter? Which could occur all year?

Discover

7. Use the **Bureau of Meteorology** weblink in the Resources tab to find the average monthly temperatures for your location. Present the statistics for the minimum and maximum temperature in a graph. Compare your graph to the one provided for Dawson. Write three clear statements comparing the average monthly temperatures of Dawson and your location. Write three clear statements comparing the monthly rainfall for Dawson and your location.

Predict

8. Which months of the year would be most popular for tourists visiting Dawson City? Provide at least two reasons to support your answer.
9. How do you think the climate would make school life in Dawson different from your life?

Think

10. What do you think happened to the early buildings when the heating from houses warmed the soil and melted the permafrost?
11. How have people tried to overcome the disadvantages of this location?

 RESOURCES — ONLINE ONLY

    Explore more with this weblink: Bureau of Meteorology

8.10 Are all settlements permanent?

Access this subtopic at **www.jacplus.com.au**

8.11 What is Old Delhi like?

Access this subtopic at **www.jacplus.com.au**

8.12 How is modern India changing?

Access this subtopic at **www.jacplus.com.au**

8.13 Review

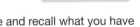

8.13.1 Review

The Review section contains a range of different questions and activities to help you revise and recall what you have learned, especially prior to a topic test.

8.13.2 Reflect

The Reflect section provides you with an opportunity to apply and extend your learning.
Access this subtopic at **www.jacplus.com.au**

TOPIC 9
Liveable places

9.1 Overview

Numerous **videos** and **interactivities** are embedded just where you need them, at the point of learning, in your learnON title at www.jacplus.com.au. They will help you to learn the content and concepts covered in this topic.

9.1.1 Introduction

Your quality of life is influenced by many factors, such as climate, landscape, community facilities, the location of your home, the sense of community identity and links to other settlements. You probably have an idea of a street, town, city or suburb where you would like to live, and your opinion may be quite different from those of others. This is because other people see different factors as important. This chapter will look at how people define and improve liveability.

Starter questions
1. What would be the good things about living in the city shown here?
2. What would be the more challenging aspects about living in this city?
3. (a) What similarities and differences are there between this *place* and where you live?
   (b) In which of these two *places* would you prefer to live and why?

The streets of Kolkata, India

9.2 What is liveability?

9.2.1 What do people think about liveability?

If you were told that Vancouver or Melbourne was the world's best place to live in, or the world's most liveable city, what would you think this means? Do city councils just brag about how good their city is or can liveability be measured? Is liveability the quality of life experienced by a city's residents?

Here are some made-up examples of what fictional people think about the liveability of their community. They come from different places and they are all trying to explain what liveability means to them.

'I think that a place is liveable if I have food every day, I do not have to walk more than 10 minutes to collect water for cooking and my father has work close by, so he is home for dinner. Liveability means warm weather, enough rain and being able to go to school every day.'
Nafula from Kenya

'I think a liveable city is a city where I can have a healthy life and where I can safely and quickly get around on foot or by bicycle, public transport or even by car — as a last resort. A liveable city is a city for everyone, including children and old people, rich and poor, and people of different religions, races and fitness levels. A liveable city should be attractive, and have good schools, a choice of things to do and fresh air.'
John from Perth

'Liveability means that I have a good job, good food, a nice house, a newish car, nice neighbours and a community that cares about my family and me.'
Oscar from western suburbs of Sydney

'Liveability is all about the **natural environment**. I think a place is liveable if the air is clean, there is plenty of water in the river and there is a healthy forest nearby. Being able to grow your own food, use renewable energy and live a simple life are all a part of what is important to me and can make a place liveable.'
Joy from Huon Valley, Tasmania

'A liveable place is somewhere I can have a computer and a television and a bed of my own in my own room. I would like a bike to get to school, three meals a day and two sisters. A liveable place would be clean, safe and modern. My grandmother and aunty would also live with us.'
Jing from a village in rural China

'Liveable cities have housing that is close to jobs, services and transport and is available for all income levels. Neighbourhoods are pedestrian-friendly with green spaces and lively retail sectors. They are mostly car-free, and have good schools and public buildings. A liveable city needs lots of different choices — choices in ways to live, places to work, shop and eat, and locations to linger in — whether alone or with other people.'
Alex, property developer from New York

'A liveable community offers many activities, celebrations and festivals that bring all of its residents together. Every year at Carnevale, my whole neighbourhood comes together to dance the samba. I would never wish to live anywhere else.'
Raul from Rio de Janeiro

'The place that I think would be the most liveable is Darwin. It has great footy grounds, public transport, good food, good houses, good shops and good schools. Where I live, my house is a dump and I cannot get anywhere unless I walk. I would like to live in Darwin and play football.'
Sam from near Alice Springs

'The community is what makes a place liveable. Being connected with my neighbours, through the community gardens, food co-op, volunteer network at our kids' school and the car-share scheme all make me feel a valued member of my community. I like knowing people who care and that we all care for each other.'
Laura from Bristol, United Kingdom

9.2 Activities

To answer questions online and to receive **immediate feedback** and **sample responses** for every question, go to your learnON title at www.jacplus.com.au. *Note:* Question numbers may vary slightly.

Explain

1. Write a statement, similar to those in this section, about the community that you live in and what makes it liveable.
2. Would the statement of liveability in your community be different if you were blind, unemployed, elderly or unable to speak English? Write what you think a community liveability statement would be for two such residents of your community.

Discover

3. Carefully read the different opinions about what makes a *place* liveable.
   (a) Make a list of the common themes mentioned by these people.
   (b) Is there a shared common definition of what makes a *place* liveable? Discuss the differences with your class.
   (c) On a map of the world, show the location of each *place* mentioned in this section.
   (d) Does the *place* in which each person lives appear to influence their definition of the term *liveability*? Discuss this with your class.
4. (a) Ask a much older person to describe the living conditions in the community they lived in as a teenager. Record or write down their memories.
   (b) Do you think the current liveability of your community is better than that described by the older person? Provide an example to support your view.

Predict

5. (a) Think about your community 50 years from now. How will the characteristics of this *place* be different? For example, think about the type of houses, the **distribution** of houses, the amount and type of traffic, the age of the population, the community facilities and other characteristics you think will be significant.
   (b) What type of inventions might improve liveability?

Think

6. We have so far described liveability in a general sense. Some times living conditions can change quite quickly. Provide examples of how natural events, political events or economic events can influence the living conditions.
7. (a) Which viewpoint or viewpoints about liveability do you agree with more than others? Give reasons for your answer.
   (b) As a class, discuss your answer to question 7(a), and see if there are any patterns within the group.

9.3 Where are the most liveable cities?

9.3.1 What is liveability?

Everyone likes to be able to tell you they are the best, or in the top 10 of some category. Cities are no different. If you look at the official websites for many international cities, they will tell you that they are the safest, wealthiest, fastest-growing or have the best events calendar. Being able to boast that a city is the world's most liveable is great publicity.

Liveability can be defined as 'the features that create a place that people want to live in and are happy to live in'. It is usually measured by factors such as safety, health, comfort, community facilities and freedom.

9.3.2 Who says which is the most liveable?

Several international organisations have created lists of the world's most liveable cities. These organisations each compare data and produce a table that ranks the liveability of cities. This information is collected for workers considering overseas transfers or for companies that may need to compensate workers who are transferred to a low-ranked city. The figures can also be used to attract migrants or investment. The various rankings compare a large number of cities; however, not all cities in the world are included in each survey.

The criteria used to produce the rankings include:

- stability or personal safety (crime, terror threats and civil unrest)
- healthcare
- culture and environment (religious tolerance, corruption, climate and potential natural disasters)
- education
- infrastructure (transport, housing, energy, water and communication)
- economic stability
- recreational and sporting facilities
- availability of consumer goods (food, cars and household items).

Figure 1 shows the top 10 and bottom 10 in the global cities liveability rankings, as released by the Economist Intelligence Unit (EIU) in August 2016. These rankings are released each year, so it is possible for you to log on (use the **Economist Intelligence Unit** weblink in the Resources tab) to get the most recent update to the rankings. This survey ranks 140 cities; a score of 100 equates to the perfect or ideal city. For the past few years, Vancouver, Melbourne and Vienna have shared the top ranking as the world's most liveable city.

The map shows that many of the world's top cities have scores that are very similar. The difference in score between the top four cities is only 0.3.

There is more than one published ranking, so obviously there is more than one list of liveable cities. With slightly different selection criteria, Zurich, Geneva and Frankfurt make it into the top 10 and, in another 2012 survey, Hong Kong was named the best city in the world.

FIGURE 1 The top 10 and bottom 10 in the global cities liveability rankings, as released by the Economist Intelligence Unit (EIU) in August 2016

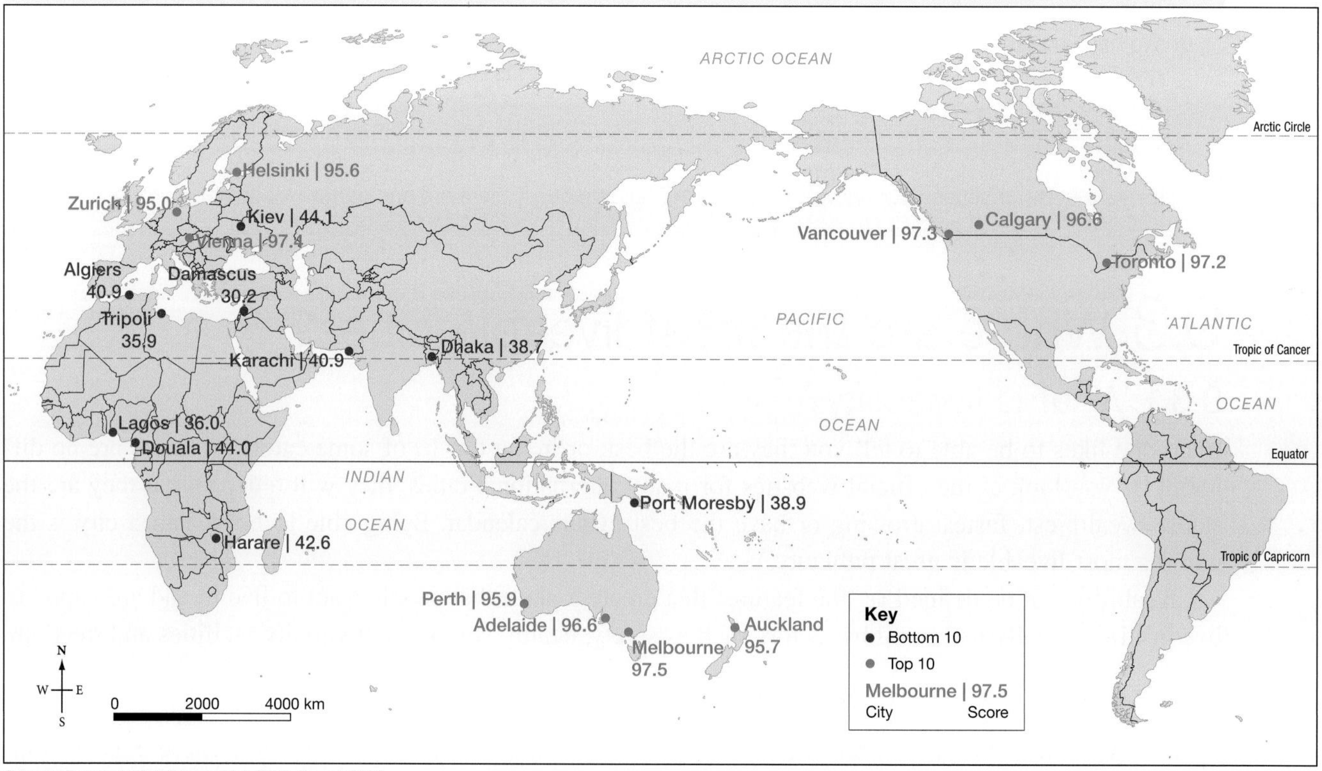

Source: Economist Intelligence Unit (EIU), August 2016

A common feature of these surveys is that cities in the United States do not rank highly even though they are very popular locations for business, travel and residence. For example, Honolulu is ranked highest at 19 while other well known cities such as New York, Los Angeles and San Francisco are ranked at about 50.

What do these top 10 liveable cities have in common?

Looking at the locations of the most liveable cities, you can see most are found in Australia, Canada or Europe. They are all mid-sized cities, have quite low **population density**, low crime rates and infrastructure that copes quite well with the needs of the local community. They are found in places where there is a **temperate climate**, perhaps with the exception of Toronto, Calgary and Helsinki, which do have very cold winters.

The top cities also tend to be modern cities, not much more than 300 years old. They have been planned so that people can travel around them by both public and private transport. They are also found in some of the world's wealthiest or most developed nations.

Australian and Canadian cities perform better than cities in the United States due to US cities' higher crime and congestion rates.

9.3 Activities

To answer questions online and to receive **immediate feedback** and **sample responses** for every question, go to your learnON title at www.jacplus.com.au. *Note:* Question numbers may vary slightly.

Remember

1. How many cities are ranked in the EIU liveability ranking?
2. What is the difference between the score of the top city (Melbourne) and the tenth city?
3. Name the three lowest ranked cities in the 2015 liveability ranking.
4. In which type of climatic region are most of the liveable cities?

Explain

5. Analyse the information in figure 1.
   (a) How many of the top 10 most liveable cities are found on each continent?
   (b) How many of the most liveable cities are found in the northern hemisphere?
   (c) Describe the distribution of the least liveable cities in the world.
   (d) How many of the least liveable cities are found on each continent?
   (e) How many of the least liveable cities are found in each hemisphere?

Discover

6. (a) Work with a partner or in a group to find the most recent population figures for each of the cities shown on the map in figure 1. List your findings. Write one sentence to describe the population of the most liveable cities. Write one sentence to describe the population of the least liveable cities.
   (b) Draw up a table or use a spreadsheet to collect at least five sets of information to compare the top 10 and bottom 10 in the liveable cities ranking. Use the population data you collected for the previous question as your first set of information. Other possible data sets are number of universities, number of hospitals, population density, any recent violence, traffic issues, the availability of public transport, housing types, presence of slums and water supply and sanitation. Comment on the differences between the most liveable and least liveable cities. Write at least three sentences.

Think

7. London and New York have a similar ranking. Why do you think these well known cities are ranked so low?
8. Why might a city suddenly fall down the liveability rankings?
9. What do you think could be done to improve a city's liveability ranking?

 my World Atlas Deepen your understanding of this topic with related case studies and questions.
○ **Polluted cities**

9.4 SkillBuilder: Drawing a climate graph

WHAT ARE CLIMATE GRAPHS?

Climate graphs, or climographs, are graphs that show climate data for a particular place over a 12-month period. They combine a column graph and a line graph. The line graph shows average monthly temperature, and the column graph shoctws average monthly precipitation (rainfall).

FIGURE 1 Climate graph for Mount Isa, Queensland

Go online to access:

- a clear step-by-step explanation to help you master the skill
- a model of what you are aiming for
- a checklist of key aspects of the skill
- a series of questions to help you apply the skill and to check your understanding.

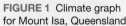
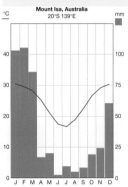

Source: Australian Bureau of Meteorology

9.5 What makes Melbourne the world's most liveable city?

Access this subtopic at **www.jacplus.com.au**

9.6 Is being the most liveable city sustainable?

9.6.1 Sustainability

Australia's major cities consistently rate among the most liveable. Liveability, however, is not always the same as sustainability.

Sustainability considers how well a community is currently meeting the needs and expectations of its population and how well it will be able to continue providing for its population.

Indicators that a place is sustainable include:

- low working hours to meet basic needs
- easy access to education
- satisfactory and affordable housing
- plenty of recycling and composting
- reliable transport
- low emissions and high air quality
- **biodiversity**
- high renewable energy use and low non renewable energy use
- good water, forests and marine health
- ability to respond to disasters.

Sustainable cities index

This annual index considers 50 leading cities and ranks each against a range of indicators. These are organised under the headings of people (society), planet (environment) and profit (economy).

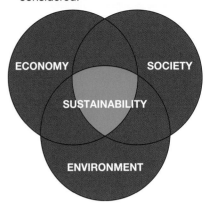

FIGURE 1 To achieve sustainability, a city's environmental, economic and social aspects must all be considered.

TABLE 1 Top 10 sustainable cities by each indicator, 2015

Ranking	People	Planet	Profit
1	Rotterdam	Frankfurt	Frankfurt
2	Seoul	Berlin	London
3	London	Copenhagen	Hong Kong
4	Sydney	Madrid	Amsterdam
5	Copenhagen	Rotterdam	Melbourne
6	Hong Kong	Amsterdam	Seoul
7	Amsterdam	Singapore	San Francisco
8	Melbourne	Rome	Brussels
9	Frankfurt	Toronto	Singapore
10	Berlin	Birmingham	Madrid

Source: ARCADIS, Sustainable Cities Index 2016

Ecological footprint

Everything we do and consume has an impact on the environment. Land is cleared to grow plants and animals; fish are caught in the sea; water is diverted for homes, businesses and farms; and most transport is powered by non-renewable resources. An **ecological footprint** calculates the land area (hectares) that would be needed to sustain an individual (expressed as per capita).

Generally, if you live in a high income country such as Australia, you are likely to have an ecological footprint that is much larger than a person who lives in a low income country such as Chad. The average ecological footprint of all people on Earth is 2.18 hectares. The average Australian footprint is about 6.8 hectares. To enjoy a sustainable way of life, the population needs to stay within the Earth's carrying capacity, and the average footprint should not be more than 1.89 hectares. As more countries develop industries and improve their standard of living, clever responses will be needed to ensure that everyone can enjoy a high standard of liveability.

Government policy can influence the ecological footprint through power generation, transport, water, industry support, rubbish collection and building regulations. Individuals can influence the ecological footprint through what they eat and buy, how they use water and power, whether they recycle and compost, and how they build their houses and travel.

FIGURE 2 Top 10 countries with the biggest and smallest ecological footprints (hectares per capita) per person, 2012

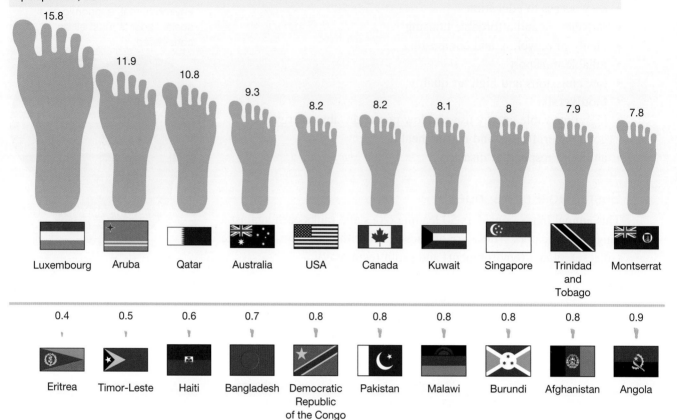

9.6 Activities

To answer questions online and to receive **immediate feedback** and **sample responses** for every question, go to your learnON title at www.jacplus.com.au. *Note:* Question numbers may vary slightly.

Remember

1. What are the three aspects that are considered in a definition of *sustainability*?
2. Refer to table 1 and your atlas. Answer the following questions.
   (a) There are 17 cities in the table. How many are located in the continent of Europe?
   (b) Which other continents are represented?
   (c) Which cities are in the top 10 for each of the three indicators for a sustainable city?
3. Refer to figure 2 and locate and describe the distribution of countries with an ecological footprint of seven or more hectares per capita. Refer to pattern, directions, continents and latitude.

Discover

4. (a) Find an image that shows living conditions in a country with an ecological footprint of over seven hectares per capita. Refer to figure 2 for possible countries.
   (b) Find an image that shows living conditions in a country with an ecological footprint of less than one hectare per capita. Refer to figure 2 for possible countries.
   (c) Annotate your images to explain how the living conditions have an impact on the ecological footprint.

Predict

5. What do you think will happen to the global ecological footprint if liveability improves on every continent?

my**World**Atlas

Deepen your understanding of this topic with related case studies and questions.
- ○ Christie Walk
- ○ Mawson Lakes

9.7 Port Moresby — a less liveable city?

9.7.1 Port Moresby

The United Nations measures people's quality of life using the Human Development Index (HDI). In 2000, Papua New Guinea was ranked 133 in the world; in 2014 its ranking had dropped to 158 (out of 188). Its largest city, Port Moresby, faces many challenges to meet the needs of its people and improve the standard of living.

Environment

Port Moresby, the capital of Papua New Guinea (PNG), is locate d on the south-eastern coastline. Its population is approximately 350 000, and it is the largest city in PNG.

Safety

The crime rate in Port Moresby is very high, and the city has a reputation as one of the most dangerous in the world. Crimes are often very violent, and gang-based crime is common. There are not enough police, and many crimes are never solved. Travellers are advised to be very careful, to not wear obviously expensive jewellery, and to avoid travelling at night.

Health

The government in Papua New Guinea spends little on preventative measures such as clean water. It also spends little on health-care. For instance, not all pregnant women can give birth in a hospital, which leads to many complications in childbirth.

FIGURE 1 Location map of Port Moresby

Source: Spatial Vision

Education

School facilities in PNG are quite poor, and attendance rates are very low, particularly for girls. Poor bus services, lack of interest and inability to pay school fees all influence the attendance rate. Only a small proportion of students complete Year 12. The **literacy rate** of 55 per cent is quite low by world standards.

Economy

The government in PNG provides no welfare. Fortunately, many families can take advantage of the good growing conditions to produce food to eat and sell. Unemployment is very high, and most work is found in the **informal sector**. Many businesses in this sector involve selling food and other goods. About half the population lives on less than $1 a day.

Life is difficult for girls, and there is much discrimination. Girls do not all get access to school; their literacy rate is lower than that of boys; child-bearing begins at a young age; and the level of violence against women is among the highest in the world.

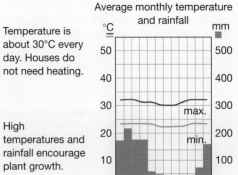

FIGURE 2 Climate graph for Port Moresby

Average monthly temperature and rainfall

Temperature is about 30°C every day. Houses do not need heating.

High temperatures and rainfall encourage plant growth.

Approximately 1000 mm of rain per year

Rainy season can cause landslides.

A distinct dry season can cause water shortages.

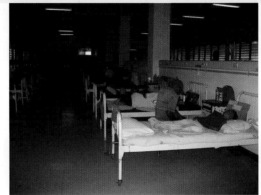

FIGURE 3 Sick children in a ward in Port Moresby General Hospital

Low spending on equipment

Shortage of medicine

Not enough doctors and nurses

Treatable diseases are common

Average life expectancy is about 65 years

Highest HIV/AIDS infection rate in the Pacific region

Infrastructure

FIGURE 4 Port Moresby is a mixture of high-rise urbanised landscapes and village landscapes.

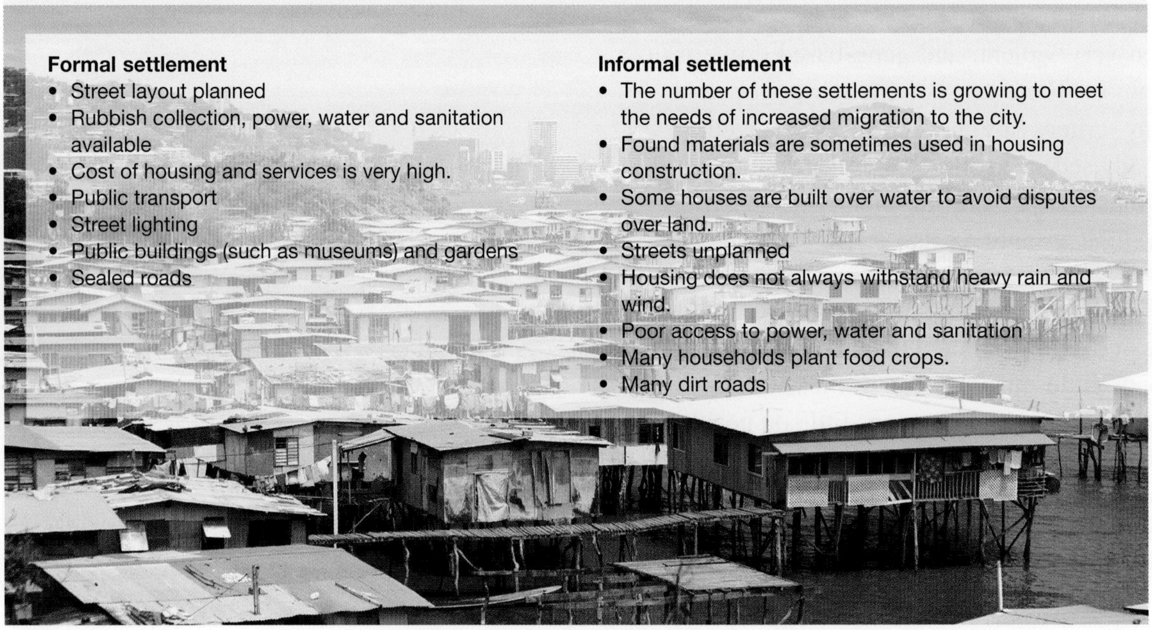

Formal settlement
- Street layout planned
- Rubbish collection, power, water and sanitation available
- Cost of housing and services is very high.
- Public transport
- Street lighting
- Public buildings (such as museums) and gardens
- Sealed roads

Informal settlement
- The number of these settlements is growing to meet the needs of increased migration to the city.
- Found materials are sometimes used in housing construction.
- Some houses are built over water to avoid disputes over land.
- Streets unplanned
- Housing does not always withstand heavy rain and wind.
- Poor access to power, water and sanitation
- Many households plant food crops.
- Many dirt roads

9.7 Activities

To answer questions online and to receive **immediate feedback** and **sample responses** for every question, go to your learnON title at www.jacplus.com.au. *Note:* Question numbers may vary slightly.

Remember

1. Refer to figure 1. At what latitude is Port Moresby?
2. Why don't all children attend school?
3. In which sector of the economy do most people find work?

Explain

4. How does environmental quality (such as climate) influence living conditions in Port Moresby?
5. Why are travellers advised to be careful in Port Moresby?
6. Refer to figure 3. Which is the biggest health issue facing Port Moresby? Why?
7. Use the **Slum life** weblink in the Resources tab to watch a video showing life in slums. Explain how slums are part of the process of city growth.

Discover

8. What is the difference between the population of Port Moresby and the biggest city in your state or territory?
9. (a) Compare the literacy rate in Papua New Guinea and Australia.
   (b) Compare the life expectancy in Papua New Guinea and Australia.
   (c) Compare the HDI ranking of Papua New Guinea and Australia.

Think

10. Find an image of an informal settlement in a country other than Papua New Guinea. What are the advantages and disadvantages of informal settlements?

learn on RESOURCES — ONLINE ONLY

✦ **Try out this interactivity:** Environmental quality (int-3096)

🔗 **Explore more with this weblink:** Slum life

9.8 Dhaka — a less liveable city?

Access this subtopic at **www.jacplus.com.au**

9.9 Is there enough to eat?

9.9.1 Distribution of hunger

A basic human requirement is food, and access to enough food is a strong measure of liveability. Even in a world where there is plenty of food and millions of people are overweight, about one person in eight does not have enough to eat.

There are approximately 870 million undernourished people in the world today. Many children in poorer countries are underweight and do not get enough food to be healthy and active.

Three-quarters of all hungry people live in rural areas, mainly in the villages of Asia and Africa (see figure 1). Most of these people depend on **agriculture** for their food. They rarely have other sources of income or employment. As a result, they may be forced to live on one quarter of the recommended calorie intake and a small amount of water each day.

If enough rain does not fall at the right time of year, crops will not grow well and there will be little grass for **livestock**. However, rainfall is not the only factor contributing to hunger. Figure 2 summarises causes of hunger.

FIGURE 1 Distribution of hunger, 2010–2012

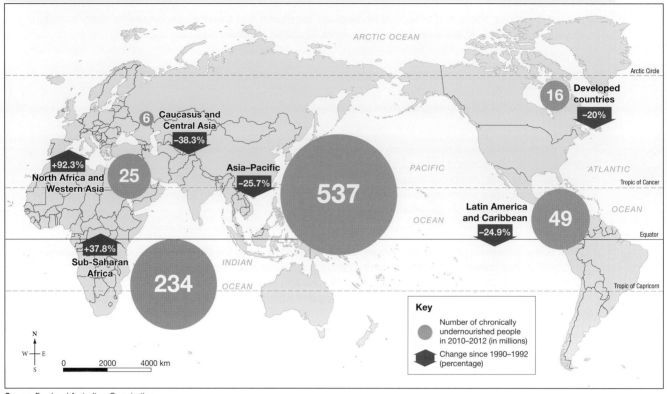

ARCTIC OCEAN

Arctic Circle

6 Caucasus and Central Asia
−38.3%

16 Developed countries
−20%

+92.3%
North Africa and Western Asia **25**

Asia–Pacific
−25.7%

537

PACIFIC

ATLANTIC

Tropic of Cancer

OCEAN

Latin America and Caribbean
−24.9%

49

OCEAN

Equator

+37.8%
Sub-Saharan Africa

234

INDIAN

OCEAN

Tropic of Capricorn

Key

Number of chronically undernourished people in 2010–2012 (in millions)

Change since 1990–1992 (percentage)

N
W E
S

0   2000   4000 km

Source: Food and Agriculture Organisation

FIGURE 2 Causes of hunger

Conflict becomes more important than food production. Farmers have difficulty getting to and from markets, crops are destroyed, many are forced off the land, and fields and water may be contaminated.

There is a lack of key agricultural infrastructure, such as roads, warehouses and irrigation. The results include high transport costs, lack of storage facilities and unreliable water supply.

Growing export crops such as coffee, cocoa and sugar produces export income while decreasing basic food production. Food becomes expensive.

The poverty cycle means the poor cannot afford to buy or produce enough food. This leaves them hungry and weak and less able to produce more food.

Poor farming techniques, deforestation, overcropping and overgrazing reduce soil fertility.

Extreme weather events such as floods, tropical storms and long periods of drought ruin crops and infrastructure.

Hunger

9.9.2 Impact of hunger

A lack of energy and poor health caused by a lack of food are made even worse by poor nutrition.

TABLE 1 The impact of hunger is felt by individuals, families, communities, regions and whole countries.

Social impacts	Economic impacts	Environmental impacts
• People become unwell. • Many people (particularly children) die. • Fathers leave in search of work. • There is political unrest.	• Food production declines. • The population of cities grows. • Poverty increases. • The government cannot afford new infrastructure.	• Soil is overused. • Too much land is cleared. • Soil fertility and local biodiversity decline.

9.9.3 Ending hunger

There is a range of organisations that focus on reducing hunger. Sometimes food is provided for immediate consumption and sometimes projects are undertaken to increase food production in the future. Actions can happen on a range of scales:

- Individuals in any country can join groups or donate to organisations that work to reduce hunger.
- The government of the affected country can provide assistance to the poor or improve infrastructure.
- Other countries can provide financial and food aid or consider the impact of their own policies.

9.9 Activities

To answer questions online and to receive **immediate feedback** and **sample responses** for every question, go to your learnON title at www.jacplus.com.au. *Note:* Question numbers may vary slightly.

Remember

1. Refer to figure 1.
   (a) Which region has the largest number of hungry people? Name three countries in this region.
   (b) Describe the *change* in the distribution of hunger. Use the following questions to help.
      - In which regions has the number of hungry people increased?
      - By what percentage?
      - In which regions has the number of hungry people decreased?
2. Copy and complete the following sentence to make it accurate. 'Most of the world's hungry people live in _____ villages in _____ and _____.'

Explain

3. Refer to figure 1. In 1990–1992 there were about one billion people who did not get enough food. How many people suffered from hunger in 2012? Is this an increase or decrease? By how many million has it changed?
4. How can poor roads contribute to hunger?

Discover

5. Work with a partner to find an example of a project that is trying to solve the immediate issue of hunger and an example of a project that is trying to make food production *sustainable*. Describe where the project is taking place and which organisation manages the project. Create an outline of the project. Refer to figure 2 and explain which of the causes will be reduced by the project.

Think

6. Here is a statement that is often in news reports: 'Hunger is caused by drought'. Is this accurate? Write your answer in a paragraph. Consider figure 2 and complete the following to help plan your answer:
   - Which causes of hunger are natural factors such as the weather?
   - Which causes are the result of actions by people?
   - Is the statement 'Hunger is caused by drought' accurate?
   - What evidence will you use to support your view?
7. Consider table 1. Provide one more example for each category of impact — social, economic and *environmental*.

9.10 How can liveability be improved?

9.10.1 Sustainable Development Goals

Many countries cannot afford to provide infrastructure for their growing population. The underlying cause of very low liveability is poverty. Reducing poverty is fundamental to improving living conditions in many parts of the world.

United Nations Development Goals

The United Nations (UN) is an organisation with members from 193 countries. In 2000, 189 countries signed a pledge to free people from extreme poverty by 2015 (Millennium Development Goals 2000–2015). In 2015, a new pledge was signed with 17 goals, each with specific targets to be reached over 15 years (Sustainable Development Goals 2015–2030).

TABLE 1 UN Development Goals

Millennium Development Goals 2000–2015	Examples of achievements of MDGs	Sustainable Development Goals 2015–2030	
Eradicate extreme poverty and hunger	Less people live in extreme poverty	No poverty	Industry, innovation and infrastructure
Achieve universal primary education	Primary school enrolments have increased	Zero hunger	Reduce inequality
Promote gender equality and empower women	Many more girls are attending school	Good health and wellbeing	Sustainable cities and communities
Reduce child mortality	More babies are surviving	Quality education	Responsible consumption and production
Improve maternal health	More mothers have access to healthcare when giving birth	Gender equality	Combat climate change
Combat HIV/AIDS, malaria and other diseases	Vaccination has reduced incidence of measles	Clean water and sanitation	Conserve and use ocean resources sustainably
Ensure environmental sustainability	Safe water is available to more people	Affordable and clean energy	Protect and use earth resources sustainably
Develop a global partnership for development	Huge increase in number of people with phone and internet	Decent work and economic growth	Provide access to justice and promote peaceful societies

Australian Government and NGOs

The Australian Government recognises that we are **global citizens**, and it supports an overseas aid program through its Department of Foreign Affairs and Trade. Overseas aid helps improve outcomes in health, education, economic growth and disaster response in many locations.

The Australian Government runs projects to improve living conditions, often working with other countries or with **non-government organisations** (NGOs). NGOs also run programs on their own. Well-known NGOs include World Vision, CARE Australia and Australian Red Cross.

Small changes, big results

Simple and **appropriate technology** can make an enormous difference to people's lives in developing countries (see figure 3).

In addition, a small amount of money can sometimes create a big difference to an individual or community group. Microfinance, or microcredit, is a system of lending small amounts of money, perhaps $150. The money is used to invest in something that can generate income. A person might buy an animal for milking and breeding, equipment for basket-making, stock for a store, or materials for jewellery-making. The loan must be repaid, but at a low interest rate, and further loans can be taken out.

FIGURE 1 Countries receiving assistance from Australia

Countries receiving aid from Australia by region
- Pacific
- East Asia
- South and West Asia
- Sub-Saharan Africa and the Middle East

Note: The point labeled Sub-Saharan Africa refers to the region rather than a specific country.

MONGOLIA

SYRIA
PALESTINIAN TERRITORIES
IRAQ
AFGHANISTAN
PAKISTAN
NEPAL
BHUTAN
BANGLADESH
MYANMAR
LAOS
VIETNAM
PHILIPPINES
CAMBODIA
SRI LANKA
MALDIVES
SUB-SAHARAN AFRICA

MARSHALL ISLANDS
PALAU
KIRIBATI
NAURU

PACIFIC OCEAN

Equator

INDIAN OCEAN
INDONESIA
TIMOR-LESTE
PAPUA NEW GUINEA
TUVALU
Tokelau (NZ)
SAMOA
VANUATU
FIJI
NIUE
TONGA
COOK ISLANDS
AUSTRALIA

N
W—E
S

0   1000   2000 km

Source: Department of Foreign Affairs and Trade

FIGURE 2 Examples of projects to improve liveability (a) A child immunisation clinic on the Kokoda Track (b) Building schools and improving education in Indonesia (c; overleaf) Planting grasses in Fiji to stabilise sea banks.

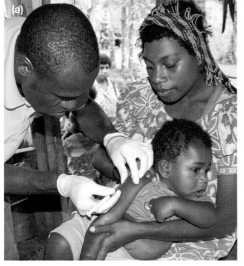

FIGURE 3 Appropriate technology (a) Electricity in Nepal is not available to all houses, so a solar lamp increases the opportunities to read. (b) In South Africa, people push hippo rollers, which make it easier to collect water from distant wells and bring it home.

9.10 Activities

To answer questions online and to receive **immediate feedback** and **sample responses** for every question, go to your learnON title at www.jacplus.com.au. *Note:* Question numbers may vary slightly.

Remember

1. Which organisation developed the Millennium Development Goals and the Sustainable Development Goals?
2. How many countries are in there in the world? What percentage of countries supported these sets of goals?

Explain

3. Study the images in figures 2 and 3. Which of the Sustainable Development Goals have been addressed in these projects?

Discover

4. Refer to figure 1. Describe the distribution of places that receive aid from Australia. Think in terms of region, such as Asia, East Asia, the Middle East, South Asia, West Asia, Pacific, Africa and the Caribbean.

5. Prepare a report about the work of one NGO involved in programs that aim to improve liveability in an overseas country. Include background information about the location of the project (country and locality); statistical data about living conditions (such as life expectancy, access to safe water, doctors per 100 000 people); and *environmental* conditions. Describe the NGO and its project and how it is aiming to improve liveability.

Think

6. Choose one of the Sustainable Development Goals. Use a visual organiser to explain how achieving this goal will improve liveability. Take into account the flow-on effects and the impact on people, the economy and the *environment*.

9.11 What makes a place liveable for you?

9.11.1 Liveability studies

A study of a region's liveability will reflect its natural characteristics and human characteristics. All communities would like a safe, healthy and pleasant place to live, a sustainable environment, the chance to earn a liveable wage, reliable infrastructure and opportunities for social interaction.

The findings of a liveability survey will be influenced by a range of factors.

- Where a person lives influences their access to services, employment and environmental features, and their address may influence their perception of the quality of the region.
- Different age groups have different views and needs.
- Current economic conditions influence a person; for example, a major employer may have closed or opened.
- Environmental conditions affect a person; for example, a region may be experiencing drought.
- Government policies influence infrastructure, housing assistance, and grants to local sports clubs.

To find out about the liveability of an area, a number of themes need to be investigated. Some of these can be gained from **census** statistics, while others can be gained only through surveys and fieldwork.

TABLE 1 Matching liveability indicators to key themes

Measure	Examples of indicators	
Social	• Population characteristics (gender, age) • Education (primary, secondary, tertiary) • Health (life expectancy, health-centre attendance, length of walking tracks, smoking rates, weight, chronic diseases) • Safety (perception, crime rates, road deaths and injuries, work safety)	• Volunteering • Voting • Aged care accommodation • Access to public transport • Membership of clubs and organisations • Diversity (ethnicity)
Environmental	• Biodiversity • Planning for the future • Water access • Waste management • Ecological footprint	• Public spaces • Household recycling • Weather • Land clearing
Economic	• Employment • Variety of businesses • Income • Financial stress • Housing types	• House ownership • Infrastructure • Internet access • Power • Car ownership

In any community there will usually be agreement about some things that improve liveability. All groups accept that safe water, sealed roads and a reliable power supply are important.

If a community wants to obtain certain kinds of items on its liveability 'wish list', it sometimes needs help from national, state or local government. Examples of such items include major roads, railways and desalination plants. Sometimes, though, a wish-list item is best obtained by an individual or community. This is the case when setting up sporting clubs, youth groups and local music events.

FIGURE 1 Community wish list: some aspects of liveability are common to all groups and some are desired by particular groups.

Community wish list

- Playgrounds
- Paths for prams
- Primary schools
- Single-person housing
- Family housing
- Friendly community
- Shopping nearby
- Paths for scooters
- Health services

- Public transport
- Neighbourhood house
- Parks and gardens
- Public seating
- Recognition of those from non-English-speaking backgrounds
- Financial security
- University of the Third Age

FIELDWORK TASK

Looking at your school environment

Geographers are particularly interested in:

- the location of things
- the distance between things
- the distribution patterns we can see when we produce a map
- the movement between places
- the connection between places
- the changes that happen over time.

How can you apply these concepts when finding ways of improving your school environment? Work in pairs to gather the data needed for the following fieldwork task. Each student will complete a report of findings.

Step 1: Study the distribution of resources and landscapes over space in the schoolyard.

- Obtain an outline map of the schoolyard.
- Walk around the schoolyard and identify different categories of land use. Design a key for your map and mark in the land uses on your map.

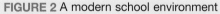
FIGURE 2 A modern school environment

- As you walk around the schoolyard, also note the landscape (slopes, swamp, bare ground, concrete and so on). Record the information about the landscape on your map. You may add to your key or annotate the map. You may also wish to use an overlay. (See the 'Creating and analysing overlay maps' SkillBuilder in subtopic 9.12.) You could also add images.
- Describe the distribution of land uses in the schoolyard. Can you identify regions? Mark these on your map. Consider using an overlay. Is there any interconnection between the landscape and land use?

Step 2: Study the patterns of movement in the schoolyard.
- Choose two places that students often walk past.
- Record the number of students who pass and the direction of travel in two 5-minute sessions.
- Add this information to your map. Which are the busiest walkways in the schoolyard?

Step 3: Make recommendations for improvements to the schoolyard.

Based on the information you have gathered, describe:
- pleasing aspects of the schoolyard
- disappointing aspects of the schoolyard
- your three most effective suggestions for improving the schoolyard.

Mark the location of the proposed changes on your map. You may include images of proposed changes.

Conduct a short survey of 10 other students to find out what they think about your three suggestions for improving the schoolyard.

When conducting your survey, there are three types of questions you need to ask: those that seek an opinion, those that seek a fact and those that seek a suggestion.
- To seek opinions, try to ask for a rating. For example, *There should be more seats in the schoolyard. Do you (a) strongly disagree? (b) disagree? (c) neither agree nor disagree? (d) agree? (e) strongly agree?*
- To seek facts, ask for a structured response. For example, *How often do you use the seating by the oval? (a) Every day (b) 1–3 times a week (c) A few times a month (d) A few times a year (e) Almost never.*
- To seek suggestions, use open-ended questions. For example, *What sorts of plants do you think should be planted along the front fence?*

Create one question for each of your suggestions. Cover each type of question. Ensure you have your map with you when you ask the questions.

Collate, or gather, the survey data for each question. For each question, count the number of responses that are the same, and present the result in words, in a table or in a graph. (You might choose a bar graph, for example). There might be a clear trend in the responses or there might be variety in the responses.

To what extent do your survey findings support your suggestions? Describe the result.

9.11 Activities

To answer questions online and to receive **immediate feedback** and **sample responses** for every question, go to your learnON title at www.jacplus.com.au. *Note:* Question numbers may vary slightly.

Remember

1. What are the three themes used when investigating liveability? Why do you think these are chosen?

Explain

2. Refer to table 1 and identify two aspects that could be placed in a different theme. Justify your suggested *change*. Suggest one more indicator that should be included. Into which theme would it belong?

Discover

3. Refer to figure 1 and use an organiser like a Venn diagram to compare and contrast the liveability wish lists for young families and older people.
4. Find a local news story about a *change* to liveability in your area. Is the change economic, social or environmental? Is the *change* predicted to be positive or negative? Will the *change* be permanent?

Think

5. How could the improvement in liveability for one age group actually help the liveability of another age group? Provide an example.

9.12 SkillBuilder: Creating and analysing overlay maps

WHAT ARE OVERLAY MAPS?

An overlay map usually consists of two or more maps of the same area. A base map is overlaid with a transparent overlay, showing different information. Overlay maps allow users to see the relationships between the information on two or more maps.

Go online to access:

- a clear step-by-step explanation to help you master the skill
- a model of what you are aiming for
- a checklist of key aspects of the skill
- a series of questions to help you apply the skill and to check your understanding.

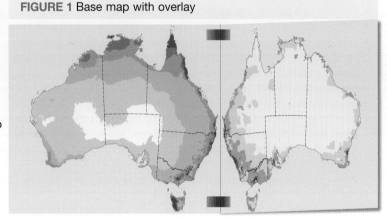

FIGURE 1 Base map with overlay

---

learn on RESOURCES — ONLINE ONLY

- Watch this eLesson: Creating and analysing overlay maps (eles-1645)
- Try out this interactivity: Creating and analysing overlay maps (int-3141)

9.13 How could my community be made more liveable?

9.13.1 Transport strategies

People in towns and cities are always looking for strategies to improve their living conditions. A community is made up of people from a range of age groups, a number of different land uses, a range of needs and a variety of interests. Ideas and plans for improvement may be overarching or targeted.

The movement of people within and between neighbourhoods is an important issue in towns and cities. The humble bicycle is now seen as a way of increasing mobility, reducing traffic congestion, reducing air pollution and boosting health. Bicycle

FIGURE 1 Recreational riding along a trail mainly designed for bicycles

tracks encourage recreational riding for all ages (see figure 1) and dedicated bicycle paths along main routes (see figure 2) encourage people to commute by bicycle, rather than car, to work and school.

In 1965, a group in Amsterdam, the Netherlands, introduced the idea of bike sharing — public bicycles that are hired, usually for short trips. This first attempt was not a success, but the idea persisted. Modern bike-sharing systems have overcome problems of theft and vandalism by using easily identifiable specialty bicycles, monitoring the bicycles' locations with radio frequency or GPS, and requiring credit-card payment or smart-card-based membership to check-out bicycles. In some places, bicycles can be located on your mobile phone, and there are more links between bicycles and existing public transport. Today there are more than 500 cities in 49 countries that have bike-sharing programs, with a combined fleet of over 500 000 bicycles. Bike-sharing programs are an example of a popular strategy that is aimed at improving liveability for a range of ages and locations within a community.

FIGURE 2 Special bicycle lanes increase the safety of bike riding

An example of a successful bike-sharing scheme is in Paris. The Vélib was introduced in 2007 and quickly doubled in size. By 2012, bicycle trips in the city had grown by 41 per cent. Bike sharing is part of a plan to reduce car traffic and pollution in Paris, which includes closing streets to cars on weekends, reducing speed limits, encouraging bus travel and extending bicycle lanes.

9.13 Activities

To answer questions online and to receive **immediate feedback** and **sample responses** for every question, go to your learnON title at www.jacplus.com.au. *Note:* Question numbers may vary slightly.

Remember

1. What are three advantages of increasing bicycle riding?
2. What problems were faced by the first bike-sharing scheme?

Explain

3. Use the **Bike-sharing** weblink in the Resources tab and scroll down the webpage to the link '(See data.)'. This will take you to an Excel document. Use this document to answer the following questions.
   (a) In 2012, how many bike-sharing programs were there in each of these regions?
   Asia–Pacific
   Europe
   Latin America
   Middle East
   North America
   (b) Name at least two countries with bike-sharing programs in each of these regions.
   (c) Which region has the most programs?
   (d) Which region has had programs for the longest period of time?
   (e) Which region has introduced programs most recently?

(f) Which region has the greatest number of bicycles? Is this the same region that has the greatest number of programs?

(g) Explain how you could present data about bike-sharing programs on a map. Provide examples of how you would show:
- countries with programs
- the number of programs per country
- the number of bicycles per country.

Discover

4. Find out about a bike-sharing scheme in Australia or overseas. Describe its location and the region it covers. Provide three other key facts about the scheme.

5. Some cities provide schemes to encourage people to ride bikes. Find out about the success of bike incentive schemes in European cities. Include the name of the city, the date of the scheme, a summary of the scheme and evidence of success or failure.

RESEARCH TASK

Teenage spaces in your community

A community is made up of a number of groups that interconnect. Teenagers are an important part of any community.

1. Produce a pie graph to show the population of your community. Refer to the latest census data provided by the Australian Bureau of Statistics. Show the percentage of each of the following categories: less than 13 years old, teenagers, adults and elderly.

Participation by teenagers in the community will be influenced by values, abilities and interests.

2. (a) Find and read a news article in the local media that is about teenagers. Note the source and date and summarise it in three dot points. For example, is it about an issue that relates to only teenagers, a space used by teenagers, an achievement by teenagers, a complaint about teenagers or a positive story about teenagers?

(b) As a class or in a small group, brainstorm a list of the ways in which teenagers participate in the community. Divide the agreed list into the following categories: informal, **formal**, social, cultural and physical.

3. (a) Find a map of your local area and use dots to show the spaces that are most attractive to teenagers. Ensure that your map satisfies all mapping conventions (BOLTSS).

(b) Describe the distribution pattern of attractive places. Is the pattern linear (in a line or lines), clustered (in small groups) or scattered?

(c) Think about the pattern you have mapped and your knowledge of the region. Are most of your favourite spaces indoors or outdoors? To what extent is there a connection between the pattern on your map and other features in your neighbourhood? Are your favourite spaces in places that are strongly influenced by the natural environment or the built environment?

(d) Add an overlay to your **base map** and use dots to show the least favourite spaces for teenagers. Describe the pattern shown on your overlay map. To what extent is there a connection between the pattern on your overlay map and other features in your neighbourhood?

4. Provide up to three examples of ways you participate in communities bigger than the local neighbourhood. For each example:

(a) Describe the scale of that community. Does it cross local council borders, state borders or national borders?

(b) Refer to a relevant map (for example, in your atlas or a street directory) to find out the direction and distance from that place to where you live. Add an arrow to your map pointing in the correct direction. Add a label to the arrow to describe the activity and the distance.

Improving your community

5. (a) Identify a space in your neighbourhood that you think could be improved for teenagers. It may be one that is currently attractive, or it may be a least favourite space.

(b) Provide an image (photograph, diagram or map) of this space. Annotate the image to describe its current characteristics.

(c) Identify the key concerns about this space. You might think about safety, tolerance, sustainability, access, inclusiveness, services, environmental quality, health and respect.

(d) How would you improve this space?
   • To help you think of suggestions, use your research skills to find out about ways in which liveability has been improved for teenagers in other parts of the world. Consider European countries in particular.
   • Discuss the ways in which the European ideas are relevant, or not relevant, to your community.
   • Provide a planning suggestion for each of the concerns you raised in question 5(c).

(e) Provide a new image to show the impact of your proposals. This could be a diagram, sketch, annotated photograph, model or whatever helps communicate what the impacts might be.

(f) Which are your two most important suggestions? What criteria did you use to choose these suggestions? Which suggestion is most likely to be implemented? Why?

(g) Compare your suggestions to the ideas of others in your class. What are the common elements? What would you put in a master plan for teenage spaces in your community?

 RESOURCES — ONLINE ONLY

 Explore more with this weblink: Bike-sharing

9.14 Review

9.14.1 Review

The Review section contains a range of different questions and activities to help you revise and recall what you have learned, especially prior to a topic test.

9.14.2 Reflect

The Reflect section provides you with an opportunity to apply and extend your learning.
Access this subtopic at **www.jacplus.com.au**

TOPIC 10
Geographical inquiry: What is my place like?

10.1 Overview

Numerous **videos** and **interactivities** are embedded just where you need them, at the point of learning, in your learnON title at www.jacplus.com.au. They will help you to learn the content and concepts covered in this topic.

10.1.1 Scenario and your task

Every person has their own idea of what their local place is like. For some people, this area can be very large; for others, it can be quite small. It really depends on where you go in your everyday life. For example, homes of relatives or friends, sports clubs, shops and parks. This means that it does not matter if your map representing your place is a different size or shape to those of friends who live in the same area. The differences simply reflect what you do and think as an individual person.

When you draw a mental map of your local place, you identify the features that you think give your neighbourhood a sense of place. All local areas have these special features that create the character or personality of the place. Many of these features can be identified on maps of the area. But there are also characteristics of your local area that you may not know about. How do you find out about these?

The Australian Bureau of Statistics (ABS) is the Commonwealth Government's organisation that has the responsibility to collect, collate and report information about Australia's people. Every five years, the ABS conducts a major survey of all Australians, as they are living on one specified day of the year. This is called

the census. The ABS then compiles this information and releases it for publication, which is when it is used by governments, businesses, companies and individuals to plan for the future.

Your task

Create a blog that presents demographic characteristics of a local place. The ABS website (www.abs.gov.au) provides a pathway for you to find out the demographic characteristics of your chosen postcode area.

10.2 Process

10.2.1 Process

- You will write your blog entries individually, but you could work in groups to share your research and create your blog.
- **Planning:** You will need to research the demographic characteristics of your chosen local place. Locate and print a map of your chosen place to accompany the data you find.

10.2.2 Collecting and recording data

- For this inquiry, use census data. Go to the ABS website (www.abs.gov.au) and select the 'Census' page. Using the 'QuickStats Search' box select the most recent census year and type your postcode or place name into the search box. This will bring a map of your local area and related data onto the screen. Choose either People, Families or Dwellings to gather information about.
- **Going further:**
  - Compare the changes to your place over time. Choose the same topic for different census years and compare the data.
  - Compare your local area with another local place in your region or city. The other place could be next to yours or a long distance away.

10.2.3 Analysing your information and data

Having collected the information about your local place, you now need to study the data and describe the patterns you have found.

1. Describe the pattern of distribution that you have produced. How does your place compare with neighbouring places?
2. What do the combinations of characteristics you have chosen tell you about the community in your postcode?
3. What does the data tell you that you did not know about the different places in the region where you live?
4. Compare and describe the changes in the data from different census years. Suggest reasons to explain these changes.
5. Visit the Resources tab and download the blog planning template to help you develop your blog. You will also see a sample blog on which you can model your own task. Use images, videos and audio files to help bring your blog to life.

10.2.4 Communicating your findings

Use an online blogging site to set up your group's blog and then enter all of the required blog entries. Be sure to create a headline for your article and add relevant tables, graphs, images, maps and videos. Your article should emphasise the important facts, and how and why they have changed over time.

10.3 Review

10.3.1 Reflecting on your work

Think back over how well you worked with your group on the various tasks for this inquiry. Determine strengths and weaknesses and recommend changes if you were to repeat the exercise. Identify one area where you were pleased with your performance, and an area where you would like to improve. Write two sentences outlining how you might be able to do this. Submit your blog and any reflection notes.

GLOSSARY

agriculture: the cultivation of land, growing of crops or raising of animals

alluvial soil: soil composed of sediments (clay, silt, sand) deposited on a floodplain by a river when it breaks its banks. This soil is rich in nutrients and is useful for agricultural production.

alluvium: the loose material brought down by a river and deposited on its bed, or on the floodplain or delta

appropriate technology: technology designed specifically for the place and the people who will use it. It is affordable and can be repaired locally.

aquifer: a body of permeable rock below the Earth's surface which contains water, known as groundwater. Water can move along an aquifer.

arid: lacking moisture; especially having insufficient rainfall to support trees or plants

artesian aquifer: an aquifer confined between impermeable layers of rock. The water in it is under pressure and will flow upward through a well or bore.

atmosphere: the layer of gases surrounding the Earth

avalanche: rapid movement of snow down a slope, usually under the influence of gravity. It can also be triggered by animals, skiers or explosions.

barometer: an instrument used to measure air pressure

base map: the map underneath an overlay

biodiversity: the variety of life in the world or in a particular habitat or ecosystem

biomass: organic (once living) matter used as fuel

blue water: the water in freshwater lakes, rivers, wetlands and aquifers

built environment: a place that has been constructed or created by people

catchment area: the area of land that contributes water to a river and its tributaries

census: an official count or survey of a population, which often seeks and records other information about people. The Australian Bureau of Statistics conducts a national census every five years.

climate change: a change in the world's climate. This can be very long term or short term, and is caused by human activity.

community: a group of people who live and work together, and generally share similar values; a group of people living in a particular region

country: the place where an Indigenous Australian comes from and where their ancestors lived; it includes the living environment and the landscape

crevasse: a deep crack in ice

cumulonimbus clouds: huge, thick clouds that produce electrical storms, heavy rain, strong winds and sometimes tornadoes. They often appear to have an anvil-shaped flat top and can stretch from near the ground to 16 kilometres above the ground.

cyclones: intense low pressure systems producing sustained wind speeds in excess of 65 km/h. They develop over tropical waters where surface water temperature is at least 26 °C.

demographic: describes statistical characteristics of a population

desalination: a process that removes salt from sea water

discharge: the volume of water that flows through a river in a given time

distribution: the way things are spread across an area

drainage basin: the entire area of land that contributes water to a river and its tributaries

Dreaming: in Aboriginal spirituality, the time when the Earth took on its present form and cycles of life and nature began; also known as the Dreamtime. Dreaming Stories pass on important knowledge, laws and beliefs.

drought: a long period of time when rainfall received is below average

ecological footprint: the total area of land that is used to produce the goods and services consumed by an individual or country

economic: relating to wealth or the production of resources

El Niño: the reversal (every few years) of the more usual direction of winds and surface currents across the Pacific Ocean. This change causes drought in Australia and heavy rain in South America

evaporate: to change liquid, such as water, into a vapour (gas) through heat

evaporation: the process by which water is converted from a liquid to a gas and thereby moves from land and surface water into the atmosphere

extensive land use: land use in which farms are huge, with few workers and not many cows or sheep per hectare

extraction: the removal of something; this may be, for example, from below the surface of the ground

famine: a situation in which there is an extreme scarcity of food

flood: inundation by water, usually when a river overflows its banks and covers surrounding land

fly in, fly out (FIFO): describes workers who fly to work in remote places, work 4-, 8- or 12-day shifts and then fly home

formal: describes an event or venue that is organised or structured

fossil fuels: fuels that come from the breakdown of living materials, and which are formed in the ground over millions of years. Examples include coal, oil and natural gas.

frostbite: damage caused to the skin when it freezes, brought about by exposure to extreme cold. Extremities such as fingers and toes are most at risk, along with exposed parts of the face.

gale force wind: wind with speeds of over 62 kilometres per hour

global citizens: people who are aware of the wider world, try to understand the values of others, and try to make the world a better place

green water: water that is stored in the soil or that stays on top of the soil or in vegetation

groundwater: a process in which water moves down from the Earth's surface into the groundwater

hailstone: an irregularly shaped ball of frozen precipitation

hailstorm: any thunderstorm that produces hailstones large enough to reach the ground

horticulture: the growing of garden crops such as fruit, vegetables, herbs and nuts

hydrologic cycle: another term for the water cycle

hypothermia: a condition in which a person's core body temperature falls below 35 °C and the body is unable to maintain key systems. There is a risk of death without treatment.

improved drinking water: drinking water that is safe for human consumption

incentive: something that encourages a person to do something

informal sector: jobs that are not officially recognised by the government as official occupations and that are not counted in government statistics

infrastructure: basic structural works needed for the operation of a modern community, such as roads, drains, bridges and the electrical supply; the facilities, services and installations needed for a society to function, such as transportation and communication systems, water and power lines

intensive agriculture: any method of farming that requires concentrated inputs of money and labour on relatively small areas of land; for example, battery hens and rice cultivation

intensive farming: farming that uses a lot of resources per hectare and changes the look of the region

intensive land use: in which farms are smaller but have more workers and machinery to produce high yields per hectare; examples are dairy and poultry farms, orchards, vegetables and feedlots

inundated: to cover with water, especially floodwater

irrigation: water provided to crops and orchards by hoses, channels, sprays or drip systems in order to supplement rainfall

isobars: lines on a map that join places with the same air pressure

literacy rate: the proportion of the population aged over 15 who can read and write

liveable city: a city that people want to live in, which is safe, well planned and prosperous and has a healthy environment

livestock: animals raised for food or other products

location: a point on the surface of the Earth where something is to be found

manufacture: to make products on a large scale

mental map: a drawing or map that contains our memory of the layout and distribution of features in a place

meteorologists: scientists who study the weather

monsoon: the rainy season in the Indian subcontinent and south-east Asia

mound spring: mound formation with water at its centre, which is formed by minerals and sediments brought up by water from artesian basins

natural disaster: an extreme event that is the result of natural processes and causes serious material damage or loss of life

natural environment: elements — such as wind, soil, flowing water, plants and animals — that influence the characteristics of an area

natural hazard: an extreme event that is the result of natural processes and has the potential to cause serious material damage and loss of life

natural resources: resources (such as landforms, minerals and vegetation) that are provided by nature rather than people

neighbourhood: a region in which people live together in a community

non-government organisations: non-profit group run by people (often volunteers) who have a common interest and perform a variety of humanitarian tasks at a local, national or international level

permafrost: permanently frozen ground not far below the surface of the soil

place: specific area of the Earth's surface that has been given meaning by people

polar vortex: a large pocket of very cold air rotating in the same direction as the Earth's orbit

population density: the number of people living in a square kilometre

precipitation: rain, sleet, hail, snow and other forms of water falling from the sky when water particles in clouds become too heavy

pull factors: positive aspects of a place; reasons that attract people to come and live in a place

push factors: reasons that encourage people to leave a place and go somewhere else

rainfall variability: the change from year to year in the amount of rainfall in a given location

rebate: a partial refund on something that has already been paid for

region: any area of varying size that has one or more characteristics in common

relative humidity: the amount of moisture in the air

remote: describes a place that is distant from major population centres

ripple effect: the flow-on effect of a particular action

riverine environment: the environment around a river or river bank

run-off: precipitation not absorbed by soil, and which runs over the land and into streams

sea change: the act of leaving a fast-paced urban life for a more relaxing lifestyle in a small coastal town

soak: place where groundwater moves up to the surface

southern oscillation: a major air pressure shift between the Asian and east Pacific regions. Its most common extremes are El Niño events

sparse: thinly scattered or unevenly distributed; often used when referring to population density

storm shelter: underground shelter where people can take refuge from a tornado

storm surge: a sudden increase in sea level as a result of storm activity and strong winds. Low-lying land may be flooded.

subsistence farmer: someone who provides food for the needs of only his or her family, leaving little or none to sell

subsistence farming: a form of agriculture that provides food for the needs of only the farmer's family, leaving little or none to sell

temperate climate: climate with generally warm summers and cool winters, without extremes

Tornado Alley: a region of the central United States, across which tornadoes are most likely to form. The core states are Texas, Oklahoma, Kansas, Nebraska, eastern South Dakota, and the Colorado Eastern Plains.

torrential rain: heavy rain often associated with storms, which can result in flash flooding

tree change: the act of leaving a fast-paced urban life for a more relaxing lifestyle in a small country town, in the bush, or on the land as a farmer

tropical depression: an area of intense low pressure, often associated with storm activity, with the capacity to develop into a tropical cyclone

tropical disturbance: a mass of storms that have the potential to develop into a typhoon

troposphere: the layer of the atmosphere closest to the Earth. It extends about 17 kilometres above the Earth's surface, but is thicker at the tropics and thinner at the poles, and is where weather occurs.

turbine: a machine for producing power, in which a wheel or rotor is made to revolve by a fast-moving flow of water, steam or air

typhoon: the name given to cyclones in the Asian region

uranium: radioactive metal used as a fuel in nuclear reactors

urban decay: situation in which a city area has fallen into a state of disrepair through its people leaving the area or not having enough resources to look after them

virtual water: all the water used to produce goods and services. Food production uses more water than any other production.

vulnerability: the state of being without protection and open to harm

water footprint: the total volume of fresh water that is used to produce the goods and services consumed by an individual or country

water scarcity: a situation that occurs when the demand for water is greater than the supply available

water stress: a situation that occurs in a country with less than 1000 cubic metres of renewable fresh water per person

water vapour: water in its gaseous form, formed as a result of evaporation

weir: a barrier across a river, similar to a dam, which causes water to pool behind it. Water is still able to flow over the top of the weir

whiteout: a weather condition where visibility and contrast is reduced by snow. Individuals become disoriented as they cannot distinguish the ground from the sky.

wilderness: a natural place that has been almost untouched or unchanged by the actions of people

INDEX

weather
  changes in   76–7
  difference from climate   77
weather maps
  reading   79
  strength of wind   105–6
weirs   97
Western Corridor Recycled
    Project   88

whiteouts   128
wilderness   162
wind
  air pressure and   103–4
  gale force wind   114
  strength and speed of   105–6
  symbols on weather maps   105–6
wind power   27
wind roses   106, 108

World Heritage sites, in
    Australia   22, 23
Y
Yukon Quest dog-sled race   181